GAS DYNAMICS

S.I. UNITS

R.K. PUROHIT
Associate Professor
Department of Mechanical Engineering
J.N.V. University,
JODHPUR - 342 001 (Raj.)

SCIENTIFIC PUBLISHERS (INDIA)
P.B. NO. 91
JODHPUR

Published by :
PAWAN KUMAR
Scientific Publishers (India)
5-A, New Pali Road, P.B. No. 91
JODHPUR -342 001 (Raj.)
FAX : 0291-2512580

Published 2006

ISBN : 81-7233-173-8

Printed at Vishal Print Process,
Delhi.

PREFACE

"Gas Dynamics" is the third publication in the series of Engineering books written so far by the author. Earlier publications i.e. Thermal Engineering, Fundamentals of Fluid Mechanics have been received well by the teachers and students. The reasons, as gathered by the feedback, are the simplicity and lucidity of the language and a lot of solved problems followed by equally numerous unsolved examples for independent practice by the students.

The experience gained thus far through previous publications have been fruitfully utilized in preparing the manuscript for this publication. Naturally, the beginners feel attracted if tough concepts are explained in simple language through practical problems drawn from day-to-day experiences of the students. Hence, as many as 113 illustrative examples have been included in the book followed by 96 unsolved problems to be tackled by the students independently. Besides it, 111 theory questions have also been provided in order to embolden the students to face the university examinations with full confidence.

Needless to emphasise, this book has been designed as a self contained learning capsule. To achieve this goal the material has been organised in a logical order and the topics have been discussed in a systematic way.

It is believed, this book mainly meant for the under graduate and post-graduate students, will captivate the interest of the teachers and students.

Manju, my wife and Abhishek and Kushboo my children have extended their co-operation and help as I was busy preparing the manuscript. Hence I gratefully acknowledge their contribution.

My indebtness is due to Prof. Dr. M.L. Mathur Ex- Vice-Chancellor who gave me guidance and suggestions. Last but not the least I express my humble regards to my father Shri B.K. Purohit, Retired Prof. of Structural Engineering who helped and guided me in producing this book. My mother Smt. Shyam Pyari also needs mention as she always encouraged and supported me in a creative endeavour like writing a book.

21st October, 1997

R.K. Purohit

CONTENTS

1

INTRODUCTION

1.1 INTRODUCTION

Any study of fluid mechanics would, however, be incomplete without a brief introduction to compressible flow behaviour. Fluid compressibility is a very important consideration in numerous engineering applications of fluid mechanics. For example, high speed subsonic, transonic, supersonic and hypersonic wind tunnels, Internal flows through rocket and gas turbine engines, external flow over moder airplanes designed to cruise faster than 0.3 of the speed of sound, and the flow inside the common internal combusion reciprocating engine.

For simplicity, in this introductory study, the following assumptions will be used

(1) Flow is steady,

(2) Flow is one dimensional

(3) Flow is non viscous.

(4) Gases will be considered perfect ($P = \rho RT$) and c_p = Constant.

(5) Macroscopic flow is considered.

Compressible flow is defined as variable density flow. The variation of fluid density for compressible flows requires attention to density and other fluid property relationships. The fluid equation of state, often unimportant for incompressible flows, is vital in the analysis of compressible flows. Also, temperature variations for compressible flows are usually significant and thus the energy equation is important.

In this chapter, one dimensional flow refers to flow involving uniform distribution of fluid properties over any flow cross sectional area. Both frictionless ($\mu = o$) and frictional ($\mu \neq o$) compressible flows are considered. We focus our attention on the compressible flow of a gas because such flows occur often. We limit are discussion to ideal gases, since the equation of state for an ideal gas is uncomplicated, yet representative of actual gas at pressures and temperatures of engineering interest, and because the flow trends associated with an ideal gas are generally applicable to other compressible fluids.

The basic equations of fluid mechanics, continuity, momentum and energy equations together with thermodynamic relationships for perfect gases, the isentropic relation p/ρ^k = constant, and the second law of thermodynamics will enable us to make a quantitative study of compressible gas flow.

1.2 IDEAL GAS RELATIONSHIPS

Any equation that relates the pressure, temperature, and specific volume or density of a substance is called an equation of state.

$$pv = RT$$

or, $$p = \rho RT \tag{1.1}$$

where R is called, the gas constant. Equation (1.1) is called the ideal gas equation of state, or simply the ideal gas relation, and gas which obeys this relation is called an ideal gas.

The gas constant R is different for each gas and is determined from

$$R = \frac{R_u}{M} \tag{1.2}$$

Where R_u is the universal gas constant and M is the molar mass (also called molecular weight) of the gas. Constant R_u is the same for all substances and its value is

$$R_u = 8.314\ kJ/kgmole\ K$$

For an ideal gas, internal energy, u, is considered to be a function of temperature only. Thus, the ideal gas specific heat at constant volume, c_v can be expressed as

$$c_v = \frac{du}{dT} \tag{1.3}$$

or, $$du = c_v\, dT$$

or, $$u_2 - u_1 = \int_{T_1}^{T_2} c_v\, dT (1.4)$$

or, $$u_2 - u_1 = c_v (T_2 - T_1) \tag{1.5}$$

The fluid property enthalpy, h, is defined as

$$h = u + pv$$

or, $$h = u + \frac{p}{\rho} \tag{1.6}$$

Since for an ideal gas enthalpy is a function of temperature only then the ideal gas specific heat at constant pressure, c_p, can be expressed as

$$c_p = \frac{dh}{dT} \tag{1.7}$$

or, $dh = c_p\, dT$

or, $h_2 - h_1 = \int_{T_1}^{T_2} c_p\, dT$ (1.8)

or, $$h_2 - h_1 = c_p (T_2 - T_1) \tag{1.9}$$

From equation (1.6)

$$h = u + P/\rho$$

or, $$h = u + RT \tag{1.10}$$

Differentiating equation (1.10) leads to

$$dh = du + R\, d\, T$$

or, $$\frac{dh}{dT} = \frac{du}{dT} + R \tag{1.11}$$

or, $$c_p = c_v + R \tag{1.12}$$

Equation (1.12) indicates that the difference between c_p and c_v is constant for each ideal gas regardless of temperature. Also $c_p > c_v$. If the specific heat ratio, k, is defined as

$$k = c_p / c_v \tag{1.13}$$

then combining equation (1.12) and (1.13) leads to

$$c_p = \frac{Rk}{k - l} \tag{1.14}$$

and $$c_v = R/k - l \tag{1.15}$$

From the first law of thermodynamics

$$dQ = p\, dv + c_v\, dT$$

According to definition of change of entropy

$$ds = \frac{dQ}{T} = \frac{pdv}{T} + \frac{c_v\, dT}{T}$$

$$ds = R\frac{dv}{v} + c_v \frac{dT}{T}$$ (1.16)

The entropy change for a process is obtained by integrating this relation (1.16) between the end states

$$s_2 - s_1 = R \ln \frac{v_2}{v_1} + c_v \ln \frac{T_2}{T_1}$$

or, $$s_2 - s_1 = R \ln \frac{\rho_1}{\rho_2} + c_v \ln \frac{T_2}{T_1} \tag{1.17}$$

Also using $\dfrac{p_1}{\rho_1 T_1} = \dfrac{p_2}{\rho_2 T_2}$ in the equation (1.17)

$$s_2 - s_1 = c_p \ln \frac{\rho_1}{\rho_2} + c_v \ln \frac{p_2}{p_1} \tag{1.18}$$

and $$s_2 - s_1 = c_p \ln \frac{T_2}{T_1} - R \ln \frac{p_2}{p_1} \tag{1.19}$$

The second law of thermodynamics requires that the adiabatic and frictionless flow of any fluid results in

$$ds = 0$$

or, $$s_2 - s_1 = 0$$

Constant entropy flow is called isentropic flow. For the isentropic flow of an ideal gas with constant c_p and c_v, we get

$$c_v \ln \frac{T_2}{T_1} + R \ln \frac{\rho_1}{\rho_2} = c_p \ln \frac{T_2}{T_1} - R \ln \frac{p_2}{p_1} = 0 \tag{1.20}$$

By combining equations (1.20) with equation (1.14) and (1.15) We obtain

$$\frac{T_2}{T_1} = \left(\frac{p_2}{p_1}\right)^{\frac{k-l}{k}} = \left(\frac{\rho_2}{\rho_1}\right)^{k-1} \tag{1.21}$$

Which is a useful relationship between temperature, pressure and density for the isentropic flow of an ideal gas. From equation (1.21) we can conclude that

$$\frac{p}{\rho^k} = constant \tag{1.22}$$

For an ideal gas with constant c_p and c_v flowing isentropically.

2

ONE DIMENSIONAL COMPRESSIBLE FLOW

2.1 THE SPEED OF SOUND

In the one dimensional fluid mechanics of an infinitesimally thin, weak pressure pulse moving at the speed of sound through a fluid at rest (Fig 2.1 (a)]. Ahead of the pressure pulse, the fluid velocity is zero and the fluid pressure and density are p and ρ. Behind the pressure pulse, the fluid velocity has changed by amounts dp and dρ. We select an infinitesimally thin control volume that moves with the pressure pulse as is sketched in Fig 2.1 (a). The speed of the weak pressure pulse is considered constant and in one direction only; thus, our control volume is inertial.

For an observer moving with this control volume (Fig 2.1 (b)) it appears as if fluid is entering the control volume through surface area A with speed 'a' at pressure p and density ρ and leaving the control volume through surface area A with speed (a – dv), pressure (p + dp), and density (ρ + dρ). When the continuity equation is applied to the flow through this control volume, the expression is

$$\rho A a = (\rho + d\rho) A (a - dV)$$

$$\rho A a = \rho A a - \rho A dV + d\rho A a - d\rho A dV$$

Since (dρ) (dV) is much smaller than the other terms, we drop it from further consideration and keep.

$$\rho dV = d\rho a$$

or, $$dV = a \frac{d\rho}{\rho} \tag{2.1}$$

For Fig 2.1 (a)

and for Fig 2.1(b)

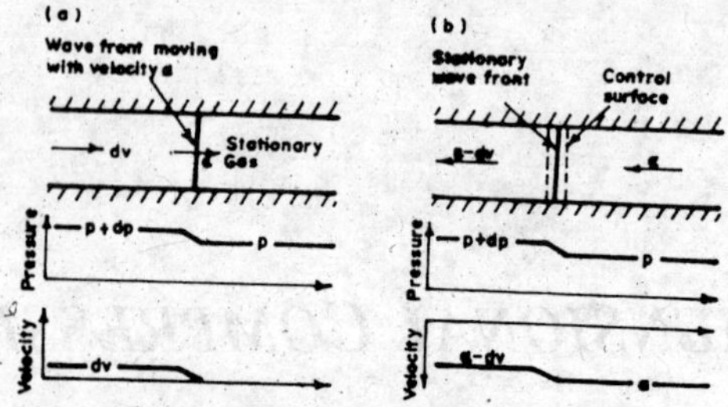

Fig 2.1 Propogation sound wave showing control volume chosen for analysis.

Applying the linear momentum equation to the flow through the control volume of the figure 2.1.

$$p\,A - (p + dp)\,A = [(a - dV) - a]\,\rho\,A\,a$$

or, $$p\,A - p\,A - dp\,A = (a - dV - a)\,\rho\,A\,a$$

or, $$dp = \rho\, a\, dV \qquad (2.2)$$

Putting the value of dv from (2.1) into (2.2)

$$dp = \rho a\left[a\frac{d\rho}{\rho}\right]$$

or, $$a^2 = \frac{dp}{d\rho}$$

or, $$a = \sqrt{\frac{dp}{d\rho}} \qquad (2.3)$$

Since the changes in pressure and temperature are extremly small, they are considered to be reversible. In addition, the temperature gradients are also small and the process is very rapid, consequnce no heat is transferred. Thus the process approaches an isentropic for which

$$\frac{p}{\rho^k} = constant$$

Taking logarithms, we obtain ln p = K ln ρ + ln constant, and differentiating

$$\frac{dp}{p} = k\frac{d\rho}{\rho}$$

or, $$\frac{dp}{d\rho} = k\frac{p}{\rho} \qquad (2.4)$$

The equation of state is

$$P = \rho\,R\,T \qquad (2.5)$$

From equation (2.4) and (2.5)

$$\frac{dp}{d\rho} = kRT \tag{2.6}$$

Substituting the above expression into equation (2.3)

$$a = \sqrt{kRT} \tag{2.7}$$

or, $$a = \sqrt{k\frac{R_u}{M}T} \tag{2.8}$$

where R_u is the universal gas constant (R_u = 8.3143 kJ/kgmole K) and M is the relative molecular mass (molecular weight) of the gas. Thus for a perfect gas the velocity of sound depends on the relative molecular mass of the gas and on its temperature.

For air, k = 1.4, R = 287 J/kg K and ambient temperature (288 K) using equation (2.7)

$$a = \sqrt{1.4 \times 287\,T} = 20.045\sqrt{T} \tag{2.9}$$

and a = 340 m/s.

The speed of sound for incompressible flow is very high as compared to compressible flow because compressible fluid may be considered to be elastic by virtue of its variable density and the speed of the sound may be thought of as a measure of elasticity of the fluid.

Since the speed of sound 'a' is uniquely related to the thermodynamic state of the fluid, it must vary from point to point in the flow just like any other property. For this reason the velocity defined above often is called the "local speed of sound".

2.2 MACH NUMBER

The Mach number, M, was introduced as a dimensionless measure of compressibility in a fluid flow. In this and subsequent sections, we develop some useful relationship involving the Mach number.

The Mach number is defined as the ratio of the value of the local flow velocity, V, to the local speed of sound, 'a'. In other words.

$$M = \frac{V}{a} \tag{2.10}$$

2.2.1 Categories of compressible flow

Flows for which M < 1 are termed subsonic, while those for which M > 1 are supersonic. Flow fields that have both subsonic and supersonic regions are termed transonic (The transonic regime occurs for Mach numbers between about 0.9 and 1.2). The transonic flows occur in aircraft compressors and fans, and

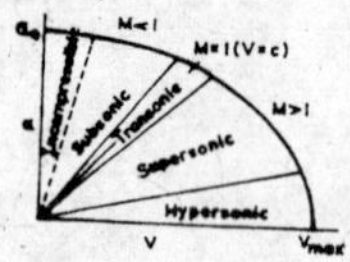

Fig 2.2 Ellipse showing ranges of compressible gas flow.

steam turbines. Moderm aircraft are mainly powered by gas turbine engines that involve transonic flows. Yet another flow regime, hypersonic flow (M > 5), is of interest in missile and re entry vehicle design or a space shuttle re enters the earth's atmosphere. Future aircraft may be expected to operate from subsonic to hypersonic flow conditions.

Fig 2.2 shows the ranges of compressible gas flow. It is also known as steady flow adiabatic eillipse.

We know that the effects of compressibility become more significant as the Mach number increases. Incompressible flows can only occur at low Mach numbers.

Consider a point source that emits weak pressure pulses. These pressure waves are spherical and expand radially outward from the point source at the speed of sound a. If a pressure wave is emitted at different times, t_{wave}, we can determine where several waves will be at a constant instant of times, t, by using the relationship

$$r = (t - t_{wave})\, a$$

where r is the radius of the sphere shaped wave emitted at time is equal to t_{wave}.

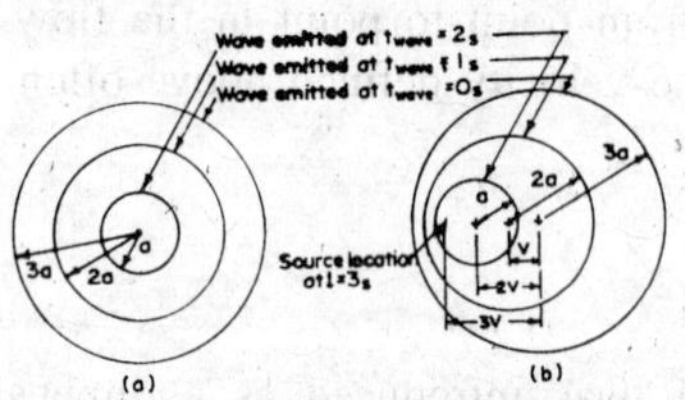

(a) Incompressible flow V = 0
(b) V < a i.e. M < 1

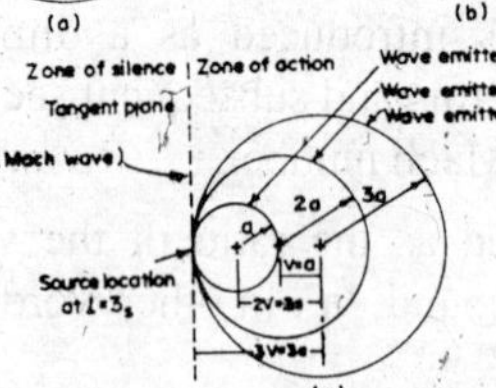

(c) V = a i.e. (M = 1)

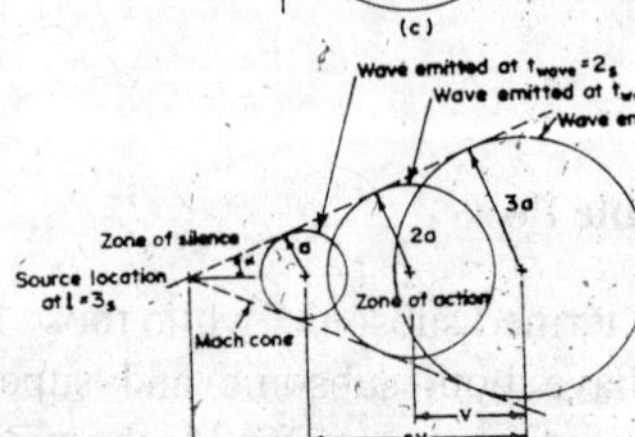

(d) V > a i.e. M > 1

Fig 2.3 Pressures waves at t = 3 s.

We are interested in determining the nature of the disturbance propagation for different speeds of the moving source. Four cases are shown in Fig 2.3.

(a) Incompressible flow

When the point source and the fluid are stationary, the pressure wave pattern is symmetrical (Fig 2.3 a) and an observer anywhere in the pressure field would hear the same sound frequency from the point source. When the velocity of the point source (or the fluid) is very small in comparison with the speed of sound; the pressure wave pattern will still be nearly symmetrical. The speed of sound in an incompressible fluid is infinitely large. Thus, the stationary point source and stationary fluid situation are representative of incompressible flows.

(b) Subsonic flow (M < 1)

When the point source moves in fluid at rest (or when fluid moves past a stationary point source) the pressure wave patterns vary in asymmetry, with the extent of asymmetry depending on the ratio of the point source (or fluid) velocity and the speed of sound. When $V/a < 1$, the wave pattern is similar to the one shown in Fig 2.3 b. This flow is considered subsonic and compressible. A stationary observer will hear a different sound frequency coming from the point source depending on where the observer is relative to the source because the wave pattern is asymmetricals. We call this phenomenon the Doppler effect. Pressure information can still travel unrestricted throughout the flow field, but not symmetrical or instantaneously.

(c) Sonic flow (M = 1)

When $M = 1$, pressure waves are not present ahead of the moving point source. The flow is sonic (Fig 2.3 (c)). If you were positioned to the left of the moving point source, you would not hear the point source until it was coincident with your location. For flow moving past a stationary point source at the speed of sound ($M = 1$), the pressure waves are all tangent to a plane that is perpendicular to the flow and that passes through the point source. The concentration of pressure waves in this tangent plane suggests the formation of a significant pressure variation across the plane. This plane is often called a Mach wave. It is to be noted that the communication of pressure information is restricted to the region of flow downstream of the Mach wave. The region of flow upstream of the Mach wave is called the zone of silence and the region of flow downstream of the tangent plane is called the zone of action.

(d) Supersonic flow (M > 1)

When $M > 1$, the flow is supersonic and the pressure wave pattern resembles the one depicted in Fig 2.3 (d). A cone (Mach cone) that is tangent to the pressure waves can be constructed to represent the Mach wave that separates the zone of silence from the zone of action in this case. The communication of pressure information is restricted to the zone of action.

2.2.2 Mach angle, Mach line and Mach cone

Mach angle cannot exist for subsonic flow (It is imaginary). The concept of Mach angle is extremely valuable in analyzing two dimensional supersonic flow.

From the Fig 2.3 d, we can see that the Mach angle, ∝, is given by

$$\sin\alpha = \frac{a}{V} = \frac{1}{M} \tag{2.11}$$

Equation (2.11) is often used to relate the Mach angle ∝, and the Mach number, M, when studying flows involving $M > 1$.

If the lines which bound the Mach cone, having an inclination of ∝, with the direction of flow, are called Mach lines.

A Mach cone that is tangent to the pressure waves can be constructed to represent the Mach wave that separates the zone of silence from the zone of action. The concentration of pressure waves at the surface of the Mach cone suggests a significant pressure, and thus density, variation across the cone surface. An abrupt density change can be visualized in a flow field by using special optics. Examples of flow visualization methods include the schlieren, shadowgraph, and interferometer techniques.

2.3 ONE DIMENSIONAL STEADY, ISENTROPIC FLOW OF AN IDEAL GAS

To define a flow, the properties of the system must be known. It should be remembered that the flow properties of a system whcih define the state of the system are the static properties and the properties which define a state at which the velocity is zero are known as stagnation properties.

It is convenient to use the stagnation state of the fluid as a reference state for compressible flow calculations. The subscript O is used to designate the stagnation state. Thus stagnation pressure and temperature are Po and To. For example, if the fluid flowing through the converging diverging duct isentropically from the atmosphere, the atmospheric, pressure and temperature would represent the stagnation state of the flowing fluid. The stagnation state can also be achieved by isentroprically decelerating a flow to zero velocity.

Mach number is such an important parameter in controlling the pattern that it will be very convenient to be able to use the fundamental equations of thermodynamics with Mach number as one of the primary variables rather than velocity. For steady flow conditions, it is relatively easy to obtain these valuable relations.

Frequently it is convenient to use the concept of stagnation enthalpy which would result when the elevation and velocity are reduced to zero.

$$h_o = h + \frac{V^2}{2} \tag{2.12}$$

The stagnation state corresponding to any fluid state is reached by proceeding from the state point to the stagnation enthalpy along a line of constant entropy. Then to reach a stagnation state, a fluid is brought to a point where the velocity and the elevation are both zero; it is brought to this condition by an isentropic process.

From equaiton (2.12)

$$h_o - h = \frac{V^2}{2} = c_p (T_o - T_p) = \frac{kR}{k-1}\left(T_o - T \right)\left[because\ c_p = \frac{kR}{k-1} \right]$$

or, $$V = \sqrt{2\,c_p (T_o - T)} = \sqrt{\frac{2k}{k-1} R\,(T_o - T)} \tag{2.13}$$

Equation (2.13) may be written as

$$T_o - T = \frac{V^2}{2}\left(\frac{k-1}{kR} \right)$$

or, $$\frac{T_o}{T} - 1 = \left(\frac{k-1}{2} \right)\left(\frac{V^2}{kRT} \right)$$

$$= \left(\frac{k-1}{2} \right)\left(\frac{V^2}{a^2} \right)$$

$$= \left(\frac{k-1}{2} \right) M^2 \qquad [because\ a = \sqrt{k\,RT}]$$

or, $$\frac{T_o}{T} = 1 + \left(\frac{k-1}{2} \right) M^2 \qquad \left[because\ M = \frac{V}{a} \right] \tag{2.14}$$

or, $$T_o = T\left[1 + \frac{k-1}{2} M^2 \right] \tag{2.15}$$

Since there must be an isentropic process between the static state and the stagnation state.

$$\frac{p_o}{p} = \left(\frac{T_o}{T} \right)^{k/k-1} = \left[1 + \frac{k-1}{2} M^2 \right]^{k/k-1} \tag{2.16}$$

$$p_o = p\left[1 + \frac{k-1}{2} M^2 \right]^{k/k-1} \tag{2.17}$$

We also know that

$$\frac{p_o}{p} = \left(\frac{\rho_o}{\rho} \right)^k$$

$$\text{or,} \quad \frac{\rho_o}{\rho} = \left(\frac{p_o}{p}\right)^{1/k} = \left[1 + \frac{k-1}{2} M^2\right]^{1/k-1} \tag{2.18}$$

$$\text{or,} \quad \rho_o = \rho\left[1 + \frac{k-1}{2} M^2\right]^{1/k-1} \tag{2.19}$$

From the above equations (2.14) to (2.19) the ratio of each isentropic stagnation property to the corresponding static property at any point in a flow field for an ideal gas can be found if the local Mach number is known. In fact, these ratios could be calculated once and for all and tabulated.

2.3.1 Three reference speeds

Since negative temperatures on the absolute scale are not attainable, it is evident from equation (2.13) that there is a maximum velocity corresponding to a given stagnation temperature. This maximum velocity, which is often used for reference purposes, is given by

$$V_{max} = \sqrt{\frac{2k}{k-1} RT_o} \tag{2.20}$$

Another useful reference velocity is the speed of sound at the stagnation temperature,

$$a_o = \sqrt{k R T_o} \tag{2.21}$$

Still a third convenient reference velocity is the critical speed, i.e., the velocity at Mach number unity. Using an asterisk to denote conditions at M = 1, we have, by definition

$$V^* \equiv a^*$$

or, using equation (2.13) and the equation for the sound speed in a perfect gas,

$$\sqrt{\frac{2k}{k-1} R (T_o - T^*)} = \sqrt{k RT^*}$$

$$\text{or,} \quad \frac{2}{k-1}\left(T_o - T^*\right) = T^*$$

$$\text{or,} \quad \frac{T^*}{T_o} = \frac{2}{k-1}\left(1 - \frac{T^*}{T_o}\right)$$

$$\text{or,} \quad \frac{T^*}{T_o}\left[1 + \frac{2}{k-1}\right] = \frac{2}{k-1}$$

$$\text{or,} \quad \frac{T^*}{T_o}\left[\frac{k-1+2}{k-1}\right] = \frac{2}{k-1}$$

or, $$\frac{T^*}{T_o}\left(\frac{k+1}{k-1}\right) = \frac{2}{k-1}$$

or, $$\frac{T^*}{T_o} = \frac{2}{k+1} \tag{2.22}$$

Substituting this value of T^* for T in equation (2.13), we get

$$V^* = a^* = \sqrt{\frac{2k}{k+1} RT_o} \tag{2.23}$$

Now, using equations (2.19) to (2.23), we get the following relations between the three reference velocities, together with the numerical values for k = 1.4;

$$\frac{a^*}{a_o} = \sqrt{\frac{2}{k+1}} = \sqrt{0.833} = 0.913 \tag{2.24}$$

$$\frac{V_{max}}{a_o} = \sqrt{\frac{2}{k-1}} = \sqrt{5} = 2.24 \tag{2.25}$$

$$\frac{V_{max}}{a^*} = \sqrt{\frac{k+1}{k-1}} = \sqrt{6} = 2.45 \tag{2.26}$$

2.3.2 Temperature, pressure and density ratios at the critical state

The particular values of the temperature, pressure, and density ratios at the critical state are found by setting M = 1 in the equations (2.14), (2.16) and (2.18). The resulting formulae, together with the numerical values for k = 1.4, are as follows :

$$\frac{T^*}{T_o} = \frac{a^{*2}}{a_o^2} = \frac{2}{k+1} = 0.833 \tag{2.27}$$

or, $$\frac{T_o}{T^*} = 1.2 \tag{2.28}$$

$$\frac{p^*}{p_o} = \left(\frac{2}{k+1}\right)^{k/k-1} = (0.8333)^{3.5} = 0.5283 \tag{2.29}$$

or, $$\frac{p_o}{p^*} = 1.8929 \tag{2.30}$$

$$\frac{\rho^*}{\rho_o} = \left(\frac{2}{k+1}\right)^{k/k-1} = (0.8333)^{2.5} = 0.6339 \quad (2.31)$$

$$\frac{\rho_o}{\rho^*} = 1.5774 \quad (2.32)$$

2.3.3 The dimensionless velocity M^*

In general the asterisk denotes the value of a property at Mach number unity, this convention is not followed in the definition of M^*.

As a dimensionless ratio, the Mach number is an extremely valuable quantity. However, it has a disadvantage because it is not a function of the velocity alone; it depends on the state of the fluid, particularly its temperature. A second disadvantage is that the Mach number may become extremely large and at high speeds it tends towards infinity. It is therefore, sometimes convenient to use a dimensionless velocity ratio defined to be the flow velocity divided by the acoustic velocity at the state where fluid velocity and sonic velocity are equal.

That is,

$$M^* = \frac{V}{a^*} = \frac{V}{V^*} \quad (2.33)$$

$$M^{*2} = \frac{V^2}{a^{*2}} = \frac{V^2}{a^2} \times \frac{a^2}{a^{*2}} = M^2 \frac{a^2}{a^{*2}} \quad (2.34)$$

The energy equation for adiabatic flow of a perfect gas is

$$h + \frac{V^2}{2} = constant$$

or, $$2\,c_p\,T + V^2 = constant$$

or, $$\frac{2\,c_p\,k\,RT}{kR} + V^2 = constant$$

or, $$\frac{2\,c_p\,a^2}{kR} + V^2 = constant$$

or, $$\frac{2\,a^2}{k-1} + V^2 = constant$$

The constant is evaluated at the three reference conditions of (i) zero speed, (ii) zero temperature, and (iii) sonic speed, we obtain three alternate and useful forms of the energy equation involving only the local velocity, local sound speed, and k

$$V^2 + \frac{2}{k-1} a^2 = \frac{2}{k-1} a_o^2 = V_{max}^2 = \frac{k+1}{k-1} a^{*2} \quad (2.35)$$

The first and last parts of equations (2.35) must be divided by a^{*2} to give

$$\frac{V^2}{a^{*2}} + \left(\frac{2}{k-1}\right) \frac{a^2}{a^{*2}} = \frac{k+1}{k-1}$$

Eliminating a^2/a^{*2} from this pair of equations

$$M^{*2} + \left(\frac{2}{k-1}\right)\frac{M^{*2}}{M^2} = \frac{k+1}{k-1}$$

or, $$M^{*2}\left[1 + \frac{2}{(k-1)M^2}\right] = \frac{k+1}{k-1}$$

or, $$M^{*2}\left[\frac{(k-1)M^2 + 2}{(k-1)M^2}\right] = \frac{k+1}{k-1}$$

or, $$M^{*2}\,[M^2(k-1) + 2] = (k-1)M^2$$

Dividing by 2

or, $$M^{*2}\left[\frac{k-1}{2}M^2 + 1\right] = \frac{(k+1)}{2}M^2$$

or, $$M^{*2} = \frac{\frac{(k+1)}{2}M^2}{1 + \left(\frac{k-1}{2}\right)M^2} \quad (2.36)$$

or, $$M^{*} = \frac{M\frac{\sqrt{(k+1)}}{2}}{\sqrt{1 + \left(\frac{k-1}{2}\right)M^2}} \quad (2.37)$$

$$M^{*2} + \left(\frac{k-1}{2}\right)M^2 M^{*2} - \frac{k+1}{2}M^2 = 0$$

or, $$\left(\frac{2}{k+1}\right)M^{*2} + \left(\frac{2}{k+1}\right)\left(\frac{k-1}{2}\right)M^2 M^{*2} - M^2 = 0$$

or, $$\left(\frac{2}{k+1}\right)M^{*2} - M^2\left[1 - \frac{k-1}{k+1}M^{*2}\right] = 0$$

or, $$M^2 = \frac{\left(\frac{2}{k+1}\right)M^{*2}}{1 - \left(\frac{k-1}{k+1}\right)M^{*2}} \quad (2.38)$$

or, $$M = \frac{M^* \sqrt{2/k+1}}{\sqrt{1-\left(\frac{k-1}{k+1}\right)M^{*2}}} \tag{2.39}$$

From equations (2.37) and (2.39) it is seen that M^* is not only a simple function of flow velocity, but is related to M in such a way that when

M < 1, then $M^* < 1$

M > 1, then $M^* > 1$

M = 1, then $M^* = 1$

M = 0, then $M^* = 0$

$M \to \infty$, then $M^* = \sqrt{\left(\frac{k+1}{k-1}\right)}$

Hence, qualitatively, M^* acts in the same fashion as M, except when M goes to à. In further discussions related to shock and expansion waves M^* will be a useful parameter because it approaches a finite number as M approaches ∞.

Fig 2.4 shows a plot of M and M^* as a function of p/p_o or isentropic flow. It is to be noted that M^* does not become infinite as p/p_o becomes very small. Also, M* is dependent on the temperature at the sonic static and not the state temperature of the fluid.

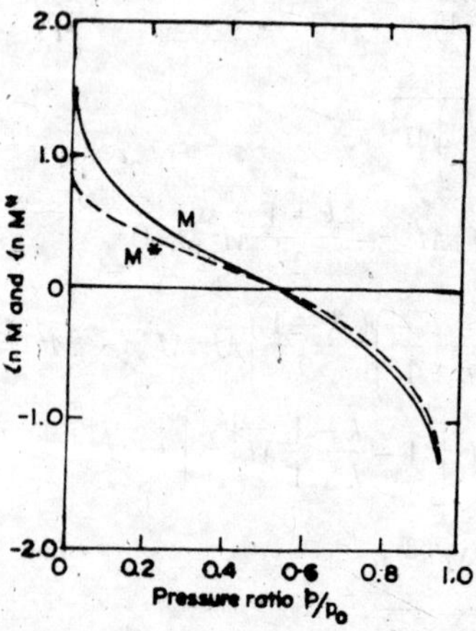

Fig 2.4 Graphical plot of M and M^* v/s p/p_o.

2.3.4 Effect of Mach number on compressibility

We can develop an estimate of a Mach number at which compressibility effects start becoming important by expanding the pressure ratio, equation (2.16), in a binomial series.

$$\frac{p_o}{p} = \left[1 + \frac{k-1}{2} M^2\right]^{k/k-1}$$

The binomial series

$$(1+x)^n = 1 + nx + \frac{n(n-1)}{2!}x^2 + \frac{n(n-1)(n-2)}{3!}x^3 + ---$$

Here $x = \frac{k-1}{2} M^2$ and $n = k/k-1$

Therefore,

$$\frac{p_o}{p} = 1 + \frac{k}{2} M^2 + \frac{k}{8} M^4 + \frac{k(2-k)}{48} M^6 + ----$$

$$\frac{p_o}{p} - 1 = \frac{k}{2} M^2 \left[1 + \frac{M^2}{4} + \frac{2-k}{24} M^4 + -----\right]$$

$$\frac{p_o - p}{p\left(\frac{k}{2} M^2\right)} = 1 + \frac{M^2}{4} + \frac{2-k}{24} M^4 + -----$$

[But p = ρ R T and $\frac{k}{2} M^2 = \frac{k}{2}\frac{V^2}{a^2} = \frac{V^2}{2RT}$

Therefore $p\left(\frac{k}{2} M^2\right) = \frac{1}{2} \rho V^2$]

or, $$\frac{p_o - p}{\rho V^2/2} = \left[1 + \frac{M^2}{4} + \frac{2-k}{24} M^4 + -----\right] \quad (2.40)$$

TABLE 2.1 Effect of compressibility

M	0.1	0.2	0.3	0.4	0.7	0.8	0.9	1
$\frac{p_o - p}{\rho V^2/2}$	1.003	1.010	1.023	1.041	1.129	1.170	1.219	1.276

Table 2.1 gives the error or deviation due to the assumption of incompressibility.

If the flow is assumed incompressible, the value of pressure coneficient (sometimes referred to as compressibility facts (C.F.)] obtained by Bernoulli's equation is unity.

$$\frac{p_o - p}{\rho V^2/2} = 1$$

Comparing the results for a fluid of constant density we see that the bracketed quantity of equation (2.40) represents the effect of compressibility Table 2.1 indicates the variation of the compressibility factor (C.F.) with M and taking k = 1.4.

It is seen that M < 0.2 compressibility affects the pressure difference by less than 1%, and the simple formula for flow at constant density is then sufficiently accurate. For larger values of M, however, the compressibility must be taken into account. At M = 1, C.F. increases by 27.6 %.

For supersonic flow, equation 2.40 is not valid because shock wave forms ahead of the pitot tube and so the fluid is not brought to rest isentropically.

It is generally accepted practice to consider a flow incompressible if its Mach number is less than 0.3.

We have seen that the proper critierion for determining whether compressibility effects are important in a flow is the magnitude of the Mach number and not whether the fluid is a gas. Moderate forces cause large enough accelerations in gases (because of their low density) for substantial density variations to occur. Using the approximate criterion of M ≈ 0.3, an atmospheric airflow with a velocity less than about 100 m/s can for most practical purposes be considered incompressible. Using the same M ≈ 0.3 criterion, a 20°C flow of water must be accelerated to a velocity of about 450 m/s before compressibility effects become important. Because of its very large mass, enormous forces are required to accelerate a water flow to this velocity. Practically, then, water flows are almost always incompressible.

ILLUSTRATIVE EXAMPLES

2.1 Standard air at rest (15°C) is accelerated isentropically. What is the Mach number when the speed becomes 500 m/s ?

Solution

We know that

$$T_o = T + \frac{V^2}{2\,c_p}$$

$$(273 + 15) = T + \frac{(500)^2}{2 \times 1005}$$

$$T = 288 - 124.378$$

$$= 163.622 \text{ K}$$

$$M = \frac{V}{a} = \frac{V}{\sqrt{k\,RT}} = \frac{500}{\sqrt{1.4 \times 287 \times 163.622}}$$

$$= \frac{500}{256.4} = 1.95 \quad \text{Ans}$$

2.2 A plane is flying at an altitude where the air temperature is 10°C. At a point on the plane where the relative air velocity is zero; the temperature is found to be 49°C. Determine the Mach number and speed of the plane.

Solution

We know that

$$\frac{T_o}{T} = \left[1 + \frac{k-1}{2} M^2\right]$$

or, $$\frac{322}{283} = 1 + 0.2\, M^2$$

or, $$M^2 = \frac{1.1378 - 1}{0.2} = 0.689$$

$\therefore$ $$M = 0.83$$

$$a = \sqrt{kRT}$$

$$= \sqrt{1.4 \times 287 \times 283} = 337.2\ m/s$$

$$V = M\,a = 0.83 \times 337.2 = 279.88$$

$$\approx 288\ m/s$$ Ans

2.3 A Jetliner is flying at M = 0.9 at an altitude of 10000 m where the air temperature and pressure are 223 K and 26.5 kN/m^2 respectively. Find the corresponding temperature and pressure on the leading edge of the wing where the air velocity relative to the wing is negligible.

Solution

(a) We know that

$$T_o = T\left[1 + \frac{k-1}{2} M^2\right]$$

$$= 223.3\left[1 + \frac{1.4-1}{2}(0.9)^2\right]$$

$$= 259.5 \text{ K}$$ Ans

$$p_o = p\left[1 + \frac{k-1}{2} M^2\right]^{k/k-1}$$

$$= 26.5\left[1 + \frac{1.4-1}{2}(0.9)^2\right]^{3.5}$$

$$= 44.82 \text{ kPa}$$ Ans

2.4 A body moves through standard air at 200 m/s. What is the pressure at a point on the body where the velocity of the air relative to the body is zero? Assume (a) compressible flow, and (b) incompressible flow. For standard air, k = 1.4, R = 287 J/kg K, T = 288 K, P = 101 kPa. and ρ = 1.225 kg/m^3

Solution

(a) Compressible flow

$$a = \sqrt{kRT}$$
$$= \sqrt{1.4 \times 287 \times 288} = 340\ m/s.$$
$$M = \frac{V}{a} = \frac{200}{340} = 0.588$$
$$\frac{p_o}{p} = \left(1 + \frac{k-1}{2} M^2\right)^{k/k-1}$$
$$= \left[1 + 0.2\,(0.588)^2\right]^{3.5}$$
$$= (1.0691)^{3.5} = 1.26$$
$$p_o = 101 \times 1.26 = 127.63\ kPa$$

Ans

(b) Incompressible flow

$$p_o = p + \frac{\rho V^2}{2}$$
$$= 101 + \frac{1.225\,(200)^2}{2 \times 1000}$$
$$= 101 + \frac{24500}{1000}$$
$$= 101 + 24.5$$
$$= 125.5 \text{ kPa}$$

Ans

2.5 Find the velocity of a bullet fired in standard air if the Mach angle is 30°.

Solution

For Standard air k = 1.4, R = 287 J/kg K, t = 15°C.

We know that

$$a = \sqrt{kRT}$$
$$= \sqrt{1.4 \times 287 \times (273 + 15)}$$
$$= \sqrt{1.4 \times 287 \times 288}$$
$$= \sqrt{115718.4}$$
$$= 340.174 \text{ m/s}$$

We also know that

$$\sin \propto = \frac{1}{M} = \frac{a}{V}$$

$$\sin 30^o = \frac{340.174}{V}$$

Therefore, V = 680.348 m/s Ans

2.6 Air at a temperature of 27°C is flowing at a Mach number of 2. Find the velocity of flow in m/s. What will be the Mach angle ? Take k = 1.4

Solution

We know that

$$a = \sqrt{kRT}$$

$$= \sqrt{1.4 \times 287 \times (273 + 27)}$$

$$= \sqrt{120540}$$

= 347.188 m/s

$$M = \frac{V}{a}$$

$$2 = \frac{V}{347.188}$$

V = 694.3774 m/s Ans

Therefore, Mach angle,

$$\sin \propto = \frac{1}{M} = \frac{1}{2} = 0.5$$

$$\therefore \quad \propto = \sin^{-1}(0.5) = 30^o$$ Ans

2.7 Through what angle must a uniform air stream with Mach number 1.5 be turned so that its static pressure is halved ?

Solution We know that

$$\frac{p_o}{p} = \left(1 + \frac{k-1}{2} M^2\right)^{k/k-1}$$

Here, k = 1.4, M = 1.5 and k/k-1 = 3.5.

Therefore,

$$\frac{p_o}{p} = \left(1 + \frac{1.4-1}{2}(1.5)^2\right)^{3.5}$$

$$= (1.45)^{3.5} = 3.671$$

In the question, p is halved i.e. p/2

Therefore,

$$\frac{p_o}{p/2} = \left(1 + \frac{k-1}{2} M^2\right)^{k/k-1}$$

$$2 \times 3.671 = \left(1 + \frac{1.4-1}{2} M^2\right)^{3.5}$$

$$(7.342)^{1/3.5} = 1 + 0.2\, M^2$$

$$1.76756 = 1 + 0.2\, M^2$$

Therefore, $M^2 = 3.8378$, $M = 1.959$

The Mach angle is defined as

$$\sin\alpha = \frac{1}{M} = \frac{1}{1.959} = 0.51$$

$\therefore \quad \alpha = \sin^{-1}(0.51) = 30.69^o$ Ans

2.8 The Static pressure to stagnation pressure ratio at a point in a flow stream is measured with a pitot static tube is being equal to 0.82. The Stagnation temperature of the fluid is 20°C. Determine the flow velocity if the fluid is air. (Take k = 1.4)

Solution

We know that

$$\frac{p_o}{p} = \left(1 + \frac{k-1}{2} M^2\right)^{k/k-1}$$

or, $$\frac{1}{0.82} = \left(1 + 0.2\, M^2\right)^{3.5}$$

or, $$1 + 0.2\, M^2 = (1.2195)^{1/3.5} = 1.05833$$

or, $$0.2\, M^2 = 0.05833$$

or, $$M^2 = 0.2916$$

$\therefore \quad M = 0.54$

Therefore,

$$\frac{T_o}{T} = 1 + \frac{k-1}{2} M^2$$

$$= 1 + 0.2\,(0.54)^2$$

$$= 1 + 0.0583 = 1.0583$$

$$\therefore \quad \frac{T}{T_o} = 0.94487$$

$$\therefore \quad T = (20 + 273)(0.94487) = 276.849 \text{ K}$$

$$a = \sqrt{kRT}$$
$$= \sqrt{1.4 \times 287 \times 276.849}$$
$$= \sqrt{111237.9282}$$
$$= 333.5235 \text{ m/s}$$

$$M = \frac{V}{a}$$

$$0.54 = \frac{V}{333.5235}$$

$$V = 180.1 \, m/s$$ Ans

2.9 An object moves at 221 m/s through standard air at sea level. What is (a) the gauge pressure and (b) the temperature at the forward stagnation point ?

Solution

The temperature and pressure at standard sea level condition are 15°C and 101.3 kPa respectively.

We know that

$$a = \sqrt{kRT}$$
$$= \sqrt{1.4 \times 287 \times 288}$$
$$= 340.17 \text{ m/s}$$

$$\therefore \quad M = \frac{V}{a} = \frac{221}{340.17} \approx 0.65$$

From the isentropic tables

At M = 0.65, $\frac{p}{p_o} = 0.753$ and $\frac{T}{T_o} = 0.922$

Therefore, $\frac{p}{p_o} = 0.753$

$$\therefore \quad p_o = \frac{101.3}{0.753} = 134.528 \, kPa \, abs$$ Ans

and $\frac{T}{T_o} = 0.922$

$\therefore$ To = 288/0.922 = 312.364 K Ans

2.10 An impact tube on an aircraft registers a stagnation pressure of 50.34 kPa gauge where the ambient static pressure is 96 kPa abs. The air temperature is 5°C.

(a) What is the free flight Mach number for the aircraft ?

(b) What is its speed ?

(c) What is the temperature on the nose of the aircraft ?

Solution

Absolute stagnation pressure = 50.34 + 96 = 146.34 k Pa

Therefore, $\frac{p}{p_o} = \frac{96}{146.34} = 0.656$

(a) From the isentropic tables

$At \; \frac{p}{p_o} = 0.656 \,, \; M = 0.8$

(b) $a = \sqrt{k\,RT}$

$= \sqrt{1.4 \times 287 \times (273 + 5)}$

$= \sqrt{1.4 \times 287 \times 288}$

$= 340.17$ m/s

V = M a = 0.8 × 340.17 = 272.14 m/s Ans

(c) From the isentropic tables, at $p/p_o = 0.656$,

$\frac{T}{T_o} = 0.886$

$\therefore \quad T_o = \frac{278}{0.886} = 313.76975\ K$ Ans

2.11 Air flow at speed of 250 m/s at temperature of 0^oC and a pressure of 600 kPa abs.

(a) What is the Mach number ?

(b) What are the stagnation pressure and temperature ?

(c) Compare the pressure rise $\left(p_o - p\right)$ with the dynamic pressure $\left(\rho\, V^2/2\right)$ of the free stream.

Solution

(a) $a = \sqrt{k\,RT}$

$= \sqrt{1.4 \times 287 \times 273}$

$= 331.1969$ m/s

$M = \frac{V}{a} = \frac{250}{331.1969} = 0.7548$ Ans

(b) $\frac{p_o}{p} = \left(1 + \frac{k-1}{2} M^2\right)^{k/k-1}$

$= \left[\ 1 + 0.2\,(0.7548)^2\ \right]^{3.5}$

$= 1.45889$

$p_o = 600 \times 1.45889 = 875.334 \, kPa$ Ans

$$\frac{T_o}{T} = 1 + \frac{k-1}{2} M^2 = 1 + 0.2\,(0.7548)^2 = 1.11394$$

$T_o = 273 \times 1.11394 = 304 \, K$ Ans

(c) $p_o - p = (876 - 600) = 275 \, kPa$

$p = \rho RT$

$600 = \rho \times 287 \times 273$

$\rho = 7.65 \text{ kg/m}^3.$

$$\therefore \quad \frac{\rho V^2}{2} = \frac{7.65 \times (250)^2}{2} = 239307.73 \, Pa = 239.3 \text{ kPa}$$ Ans

2.12 A supersonic wind tunnel operates at test section Mach number, M = 1.8. Operating conditions in the tunnel test section are – 80°C and 258 kPa (gauge). A normal shock that stands in front of a total head tube reduces the stagnation pressure by 18.7 %. Calculate (a) the local isentropic stagnation conditions for the test section conditions, and (b) the pressure sensed by the total head tube. Take barometric pressure as 101 kPa.

Solution

$$\frac{T_o}{T} = 1 + \frac{k-1}{2} M^2$$

$= 1 + 0.2\,(1.8)^2 = 1.648$

$\therefore \quad T_o = T\,(1.648) = 193 \times 1.648 = 318 \, K$ Ans

$$\frac{p_o}{p} = \left(\frac{T_o}{T}\right)^{k/k-1} = (1.648)^{3.5} = 5.7458$$

$p_o = p\,(5.7458)$

$= (258 + 101)\,(5.7458)$

$= 359 \times 5.7458$

$= 2062.74 \text{ kPa} = 20.6274 \text{ bar}$ Ans

$p_{probe} = 0.813\, p_o = 0.813 \times 20.6274 = 16.77 \text{ bar}$ Ans

2.13 The hot gas stream at the turbine inlet of a jet engine (JT 9-D) is at 1300 K and 140 kPa. The Mach number at that location is 0.32. Calculate the critical conditions (temperature, pressure and flow speed) that correspond to these conditions. Assume the fluid properties are those of pure air.

Solution

$$\frac{T_o}{T} = 1 + \frac{k-1}{2} M^2$$

$$= 1 + 0.2\,(0.32)^2$$

$$= 1 + 0.02048$$

$$= 1.02048$$

$$T_o = T\,(1.02048)$$

$$= 1300 \times 1.02048 = 1326.624 \text{ K}$$

$$\frac{p_o}{p} = \left(\frac{T_o}{T}\right)^{k/k-1} = (1.02048)^{3.5} = 1.0735$$

$$p_o = p\,(1.0735) = 140 \times 1.0735 = 150.29\ kPa$$

At critical conditions, M = 1

$$\frac{T_o^*}{T^*} = 1.2 \therefore T^* = \frac{T_o^*}{1.2} = \frac{1326.624}{1.2} = 1105.2\ K \qquad \text{Ans}$$

$$\frac{p_o^*}{p^*} = \left(\frac{T_o^*}{T^*}\right)^{k/k-1} = (1.2)^{3.5} = 1.8929$$

$$p^* = p_o^*/1.8929 = \frac{150.29}{1.8929} = 79.395\ kPa \qquad \text{Ans}$$

$$V^* = a^* = \sqrt{k\,RT^*}$$

$$= \sqrt{1.4 \times 287 \times 1105.2} = 666.385\ m/s \qquad \text{Ans}$$

2.14 Show that for an isentropic flow

$$V_{max} = \sqrt{\frac{2a^2}{k-1} + V^2}$$

Solution

Energy equation without shaft work and heat transfer is written as

$$dh + V\,dV = 0$$

Here, $dh = c_p\,dT = \dfrac{k\,R\,dT}{k-1}$

Therefore, $\dfrac{k\,R\,dT}{k-1} + V\,dV = 0$

Integrating with limits from T_o to T and V_o to V.

$$\frac{kR}{k-1}\int_{T_o}^{T} dT + \int_{V_o}^{V} V\,dV = 0$$

or, $$\frac{kR}{k-1}(T - T_o) + \frac{V^2 - V_o^2}{2} = 0$$

For initial velocity equal to zero i.e. $V_o = 0$

Therefore,

$$\frac{kR}{k-1}(T - T_o) + \frac{V^2}{2} = 0$$

or, $$\frac{kRT}{k-1} + \frac{V^2}{2} = \frac{kRT_o}{k-1}$$

$$\frac{a^2}{k-1} + \frac{V^2}{2} = \frac{a_o^2}{k-1} \tag{1}$$

For V_{max}, a = 0, Substituting in equation (1)

$$\therefore \frac{V_{max}^2}{2} = \frac{kRT_o}{k-1} = \frac{a_o^2}{k-1}$$

or, $$V_{max}^2 = \frac{2a_o^2}{k-1} \tag{2}$$

From equation (1)

$$a_o^2 = a^2 + \frac{V^2}{2}(k-1) \tag{3}$$

Substituting the value of a_o^2 from (3) into (2), we get

$$V_{max}^2 = \frac{2}{k-1}\left[a^2 + \frac{V^2}{2}(k-1)\right]$$

$$= \frac{2a^2}{k-1} + V^2$$

$$V_{max} = \sqrt{\frac{2a^2}{k-1} + V^2}$$ Proved.

2.15 Show that for an isentropic flow

$$a^* = \sqrt{\frac{2a^2 + V^2(k-1)}{k+1}}$$

Solution We know that

$$\frac{a_o^2}{k-1} = \frac{a^2}{k-1} + \frac{V^2}{2}$$

or, $$a_o^2 = a^2 + \frac{V^2}{2}(k-1) \quad (1)$$

We know that

$$M = \frac{V}{a} \quad (2)$$

When M = 1, V = a that is critical condition

$$\therefore \quad V = a = a^*$$

Putting in the equation (1)

$$a_o^2 = a^{*2} + \frac{a^{*2}}{2}(k-1)$$

or, $$a_o^2 = a^{*2}\left[1 + \frac{k-1}{2}\right]$$

$$= a^{*2}\left[\frac{2+k-1}{2}\right]$$

$$= a^{*2}\left[\frac{k+1}{2}\right]$$

or, $$a^{*2} = \frac{2a_o^2}{k+1} \quad (3)$$

$$= \frac{2}{k+1}\left[a^2 + \frac{V^2}{2}(k-1)\right]$$

$$= \frac{2a^2}{k+1} + \frac{V^2(k-1)}{(k+1)}$$

$$a^* = \sqrt{\frac{2a^2 + V^2(k-1)}{k+1}}$$ **Proved.**

2.16 Establish the relation

$$M = \frac{V}{V_{max}}\left[\frac{2/(k-1)}{1 - \frac{V^2}{V_{max}^2}}\right]^{1/2}$$

for an ideal gas and hence prove that

$$\frac{T}{T_o} = 1 - \frac{V_2}{V_{max}^2}$$

From this show that $\frac{T}{T_o}$ is independent of the thermodynamic properties of the gas itself.

Solution

(a) We know that

$$V_{max} = \sqrt{\frac{2k}{k-1} R T_o}$$

$$V^2_{max} = \frac{2}{k-1}(kR T_o)$$

$$= \frac{2a_o^2}{k-1} \quad (1)$$

We also know that

$$\frac{a_o^2}{k-1} = \frac{a^2}{k-1} + \frac{V^2}{2}$$

or, $$a_o^2 = a^2 + \frac{V^2}{2}(k-1)$$

Putting the value of a_{o2} in equation (1)

$$V^2_{max} = \frac{2}{k-1}\left[a^2 + \frac{V^2}{2}(k-1)\right]$$

$$V^2_{max} = \frac{2a^2}{k-1} + V^2$$

or, $$V^2_{max} - V^2 = \frac{2a^2}{k-1}$$

or, $$V^2_{max}\left[1 - \frac{V^2}{V^2_{max}}\right] = \frac{2a^2}{k-1}$$

or, $$V^2_{max}\left(1 - \frac{V^2}{V^2_{max}}\right) = \frac{2}{k-1}\left(\frac{V^2}{M^2}\right)$$

or, $$M^2 = \frac{\frac{2}{k-1}\left(\frac{V^2}{V^2_{max}}\right)}{\left(1 - \frac{V^2}{V^2_{max}}\right)}$$

$$= \frac{V^2}{V^2_{max}}\left(\frac{2/(k-1)}{1 - V^2/V^2_{max}}\right)$$

$$M = \frac{V}{V_{max}} \sqrt{\frac{2/(k-1)}{1 - V^2/V_{max}^2}}$$ Proved

(b) For an ideal gas

$$\frac{T_o}{T} = 1 + \frac{k-1}{2} M^2$$

$$= 1 + \frac{k-1}{2}\left[\frac{V^2}{V_{max}^2}\frac{[2/(k-1)]}{1 - V^2/V_{max}^2}\right]$$

$$= 1 + \frac{V^2}{V_{max}^2}\left(\frac{V_{max}^2}{V_{max}^2 - V^2}\right)$$

$$\frac{T_o}{T} = 1 + \frac{V^2}{V_{max}^2 - V^2} = \frac{V_{max}^2}{V_{max}^2 - V^2}$$

$$\frac{T}{T_o} = 1 + \frac{V_{max}^2 - V^2}{V_{max}^2} = 1 - \frac{V^2}{V_{max}^2}$$ Proved.

The relation T/To is a function of velocity and maximum velocity therefore we can say that it is independent of thermodynamic property.

2.17 Air at stagnation condition has a temperature of 800 K. Determine the velocity of sound and the maximum fluid velocity. What is the velocity of sound when the flow is at half the maximum velocity ?

Solution

We know that

$$a = \sqrt{kRT}$$

$$= \sqrt{1.4 \times 287 \times 800}$$

$$= 566.956 \text{ m/s}$$ Ans

We know that

$$V_{max}^2 = \frac{2\,a_o^2}{k-1} \quad (1)$$

Therefore, $V_{max} = a_o \sqrt{\frac{2}{k-1}}$

$$= a_o \sqrt{5}$$

$$= 566.956 \times 2.24 = 1267.7539 \text{ m/s}$$ Ans

We know that

$$a_o^2 = a^2 + \frac{V^2}{2}(k-1)$$

Substituting in equation (1)

$$V_{max}^2 = \frac{2}{k-1}\left[a^2 + \frac{V^2}{2}(k-1)\right]$$

$$= \frac{2a^2}{k-1} + V^2$$

or, $$\left(V_{max}^2 - V^2\right) = \frac{2a^2}{k-1}$$

or, $$a^2 = \left(V_{max}^2 - V^2\right)\left(\frac{k-1}{2}\right)$$

Here, $V = \frac{V_{max}}{2}$

Therefore,

$$a^2 = \left(1 - \frac{1}{4}\right)\left(\frac{0.4}{2}\right)V_{max}^2$$

$$= \frac{0.75 \times 0.4}{2}(1267.7539)^2 = 0.15 \times (1267.7529)^2$$

Therefore, a = 490.9989 ≈ 491 m/s Ans

2.18 The stream of air flows at the rate of 1 kg/s in a duct of 100 mm diameter. The stagnation temperature is 310 K. At one section of duct, the static pressure is 40 kPa. Calculate the stagnation pressure, velocity and Mach number at this section. Take R = 287 J/kg K.

Solution

$$\dot{m} = \rho A V$$

$$1 = \rho \frac{\pi}{4}(0.1)^2 V$$

$$\rho V = 127.32395$$

or, $$\rho \approx \frac{127.3}{V} \qquad (1)$$

$$p = \rho RT$$

$$40 = \frac{127.3}{V} \times 0.287 \times T$$

$$\therefore \quad V = \frac{127.3 \times 0.287\, T}{40} = 0.9133\, T \qquad (2)$$

We also know that

$$V^2 = \frac{2kR}{k-1}(T_o - T)$$

$$V^2 = \frac{2 \times 1.4 \times 287}{0.4}(310 - T)$$

$$V^2 = 2009\,(310 - T)$$

$$= 2000 \times 310 - 2009\,T$$

$$= 622790 - 2009\,T \qquad (3)$$

Substituting the value of T from (2) in (3), we get,

$$V^2 = 622790 - 2009 \times \frac{V}{0.9133}$$

or, $$V^2 = 622790 - 2199.7153\,V$$

or, $$V^2 + 2199.7153\,V - 622790 = 0$$

Solving we get

$$V = 251 \text{ m/s}$$ Ans

Therefore from (2)

$$T = \frac{251}{0.9133} = 274.8275\,K \approx 275\,K$$

Therefore,

$$\frac{T_o}{T} = 1 + \frac{k-1}{2}M^2$$

$$\frac{310}{275} = 1 + 0.2\,M^2$$

$$1.12727 = 1 + 0.2\,M^2$$

$$M^2 = \frac{0.12727}{0.2} = 0.63636$$

$\therefore$ $$M = 0.7977 \approx 0.8$$ Ans

$$\frac{p_o}{p} = \left(\frac{T_o}{T}\right)^{k/k-1} = (1.2727)^{3.5} = 1.52$$

$$p_o = 1.52 \times 40 = 60.8356 \text{ kPa}$$ Ans

2.19 If at a point in a flow stream of air the velocity is 150 m/s and the local temperature is 230°C, What are (a) a, the local sonic velocity at the local temperature, (b) a^*, the sonic velocity where the flow changed isentropically to a Mach number of 1, and (c) a_o, the sonic velocity at the stagnation point ? Take R = 287 J/kg K.

Solution

(a) $a = \sqrt{kRT}$

$= \sqrt{1.4 \times 287 \times (273 + 230)}$

$= \sqrt{1.4 \times 287 \times 503}$

$= 449.561 \approx 450$ m/s   Ans

We know that

$$\frac{V^2}{2} + \frac{a^2}{k-1} = \frac{a_o^2}{k-1} = \{(k+1)\frac{a^{*2}}{2(k-1)} \qquad (1)$$

(b) From equation (1)

$$\frac{V^2}{2} + \frac{a^2}{k-1} = \frac{a_o^2}{k-1}$$

$$\frac{(150)^2}{2} + \frac{(450)^2}{1.4-1} = \frac{a_o^2}{1.4-1}$$

$$11250 + 5062.50 = \frac{a_o^2}{0.4}$$

$$a_o^2 = (517500) \times 0.4 = 207000$$

$a_o = 454.972 \approx 455\ m/s$ Ans

(c) From equation (1)

$$\frac{a_o^2}{k-1} = \{(k+1)\frac{a^{*2}}{2(k-1)}$$

$$a^{*2} = \frac{2\,a_o^2}{k+1}$$

$$= \frac{2\,(455)^2}{1.4+1}$$

$$= \frac{2}{2.4}(455)^2$$

$$a^* = \sqrt{0.833} \times 455$$

$= 415.356 \approx 416$ m/s   Ans

2.20 Air at rest at 90°C is accelerated isentropically.

(a) What is the air speed in m/s when the Mach number becomes 0.8 ?

(b) What is the air speed when the flow becomes sonic ?

(c) What is the Mach number when the air speed becomes 600 m/s ?

Solution

$$a_o = \sqrt{kRT_o}$$
$$= \sqrt{1.4 \times 287 \times (273 + 90)}$$
$$= \sqrt{145853.4}$$
$$= 381.9 \text{ m/s}$$

(a) We know that

$$\frac{T_o}{T} = 1 + \frac{k-1}{2} M^2$$

or,
$$T = \frac{T_o}{1 + 0.2\,M^2} = \frac{363}{1 + 0.2\,(0.8)^2}$$
$$= \frac{363}{1.128} = 321.8\ K$$
$$a = \sqrt{kRT} = \sqrt{1.4 \times 287 \times 321.8} = 359.586\ m/s$$

Therefore, V = M a = 0.8 × 359.586 = 287.669 m/s Ans

(b)
$$a^* = a_o \sqrt{\frac{2}{k+1}} = \frac{381.9}{\sqrt{1.2}} = 348.625\ m/s = V^*$$

(c)
$$T_o = T + \frac{V^2}{2\,c_p}$$
$$T = T_o - \frac{V^2}{2\,c_p}$$
$$= 363 - \frac{(600)^2}{2 \times 1005}$$
$$= 363 - 179.1 = 183.895 \text{ K}$$
$$M = \frac{V}{C} = \frac{600}{\sqrt{kRT}} = \frac{600}{\sqrt{1.4 \times 287 \times 183.895}}$$
$$= \frac{600}{271.825} = 2.20729$$ Ans

2.21 Show that the ratio of stagnation density ρ_o to density ρ, where the Mach number is M, for isentropic flow is

$$\frac{\rho_o}{\rho} = \left(1 + \frac{k-1}{2} M^2\right)^{1/(k-1)}$$

and by expanding in a power series, show that for very small M

$$\frac{\Delta\rho}{\rho} \approx \frac{M^2}{2}$$

(b) Suppose gas flow is defined as incompressible if the variation in density is no greater than 2 percent $\left(\rho_o/\rho \le 1.02\right)$. What is the maximum Mach Number for which the flow may then be considered incompressible ? Show that this has no significant dependency on the type of gas.

Solution

We know that

$$\frac{p_o}{\rho_o T_o} = \frac{p}{\rho T} \quad \text{and} \quad \frac{p}{\rho^k} = \frac{p_o}{\rho_o^k}$$

$$\text{or,} \quad \frac{\rho_o}{\rho} = \left(\frac{p_o}{p}\right)^{1/k} = \left(\frac{T_o}{T}\right)^{1/(k-1)}$$

$$\text{or,} \quad \frac{\rho_o}{\rho} = \left[1 + \frac{k-1}{2} M^2\right]^{1/(k-1)}$$

$$\frac{\rho_o}{\rho} = 1 + \left(\frac{1}{k-1}\right)\left(\frac{k-1}{2}\right) M^2$$

$$\approx 1 + \frac{M^2}{2} \quad \text{for M} << 1$$

$$\text{Thus} \quad \frac{\Delta \rho}{\rho} \approx \frac{M^2}{2}$$

$$\text{(b)} \quad 0.02 = \frac{M^2}{2}$$

$$M_{max} \approx \sqrt{2 \times 0.02} = 0.2$$ Ans

$$\text{For k = 1.67,} \quad \frac{2-k}{8} M^4 \approx 6.6 \times 10^{-5} << 0.02$$

$$\text{For k = 1.3,} \quad \frac{2-k}{8} M^4 \approx 1.4 \times 10^{-4} << 0.02$$

2.22 For aircraft flying at supersonic speeds, the lift and drag coefficient are functions of Mach number only. A supersonic transport with wingspan of 75 m is to fly at an air speed of 780 m/s at an altitude of 20 km on a standard day. Performance of the aircraft is to be measured from model test data using a scale model with 0.9 m wingspan in a supersonic wind tunnel. The wind tunnel is to be supplied from a large reservoir of compressed air, which can be heated if desired. The static temperature of air in the test section is to be 10°C to avoide freezing of moisture. At what air speed should the wind tunnel tests be run to duplicate the Mach number of the prototype ? What must be the stagnation temperature in a reservoir ? What pressure is required in the reservoir if the test section pressure is to be 10 kPa ?

Solution

At 20 km, T = 216.7 K

$$a = \sqrt{kRT} = \sqrt{1.4 \times 287 \times 216.7} = 295 \text{ m/s}$$

Thus for aircraft

$$M = \frac{V}{a} = \frac{780}{295} = 2.64 = M_p$$

$$a_m = \sqrt{kRT_m} = \sqrt{1.4 \times 287 \times 283} = 337\ m/s$$

Mach number of model = Mach number of prototype

$$M_m = M_p$$

$$\frac{V_m}{a_m} = 2.64$$

$$\therefore \quad V_m = 2.64 \times 337 = 890 \text{ m/s} \qquad \text{Ans}$$

$$\frac{T_o}{T} = 1 + \frac{k-1}{2}M^2$$

$$T_o = T\,[1 + 0.2\,(2.64)^2] = 283 \times 2.39392 = 677.48 \text{ K} \qquad \text{Ans}$$

$$\frac{p_o}{p} = \left(\frac{T_o}{T}\right)^{k/k-1} = (2.39392)^{3.5} = 21.226$$

$$\therefore \quad p_o = p \times 21.226 = 10 \times 21.226 = 212.26 \text{ kPa} \qquad \text{Ans}$$

2.23 Air flows steadily through a duct from 350 kPa, 60°C, and 183 m/s at the initial state to a Mach number of 1.3 at the outlet, where the local isentropic stagnation conditions are known to be 385 kPa and 77°C. Compute the isentropic stagnation pressure and temperature at the inlet and the static pressure and temperature at the duct outlet.

Solution At Section (1)

$$a_1 = \sqrt{kRT_1} = \sqrt{1.4 \times 287 \times (273 + 60)}$$

$$= 20.1\sqrt{333} = 365.786 \approx 366 \text{ m/s}$$

$$M_1 = \frac{V_1}{a_1} = \frac{183}{366} = 0.5$$

$$\frac{p_{o1}}{p_1} = \left[1 + \frac{k-1}{2}M_1^2\right]^{k/k-1} = \left[1 + 0.2\,(0.5)^2\right]^{3.5} = 1.186$$

$$\therefore \quad p_{o1} = 1.186\,p_1 = 1.186 \times 350 = 415\ kPa \qquad \text{Ans}$$

$$\frac{T_{o_1}}{T_1} = 1 + \frac{k-1}{2}M_1^2 = 1 + 0.2\,(0.5)^2 = 1.05$$

$\therefore \quad T_{o1} = 1.05\, T_1 = 1.05 \times 333 = 350\, K$ Ans

At Section (2)

$$\frac{p_{o_2}}{p2} = \left[1 + \frac{k-1}{2} M_2^2\right]^{k/k-1} = \left[1 + 0.2\,(1.3)^2\right]^{3.5} = 2.77$$

$$p_2 = \frac{p_{o2}}{2.77} = \frac{385}{2.77} \approx 139\, kPa$$ Ans

$$\frac{T_{o_2}}{T2} = 1 + 0.2\,(1.3)^2 = 1.338 \quad \therefore\ T_2 = \frac{350}{1.338} = 262\, K$$ Ans

2.24 An F-4 aircraft makes a high speed pass over an airfield on a day when the temperature is 35°C. The aircraft flies at M = 1.4 and 200 m altitude. Calculate the air speed of the aircraft. How long after it passes directly overhead does its Mach cone pass a point on the ground ?

Solution

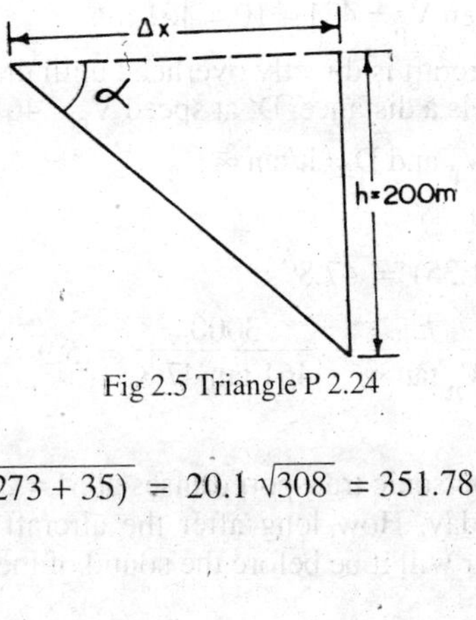

Fig 2.5 Triangle P 2.24

$$a = \sqrt{k\,RT}$$

$$= \sqrt{1.4 \times 287 \times (273 + 35)} = 20.1\sqrt{308} = 351.78 \approx 352 \text{ m/s}$$

$$M = \frac{V}{a}$$

$\therefore \quad V = M\,a = 1.4 \times 352 = 493$ m/s

For the instant the aircraft is directly over head until the Mach cone reaches the ground, the plane travels a distance Δ x at speed V = 493 m/s.

$$\sin\alpha = \frac{1}{M} = \frac{1}{1.4} = 0.7143$$

$\therefore \quad \alpha = 45.5847^\circ$

$$\tan\alpha = \frac{h}{\Delta x}$$

$$\therefore \quad \Delta x = h/\tan\alpha = \frac{200}{\tan 45.5847} = 195.96\, m$$

Since the plane moves at constant speed V.

$\Delta x = V \Delta t$ and $\Delta t = \Delta x / V = 196/493 = 0.398$ s Ans

2.25 An F-5G aircraft passes overhead at an altitude of 3 km. The aircraft flies at M = 1.35; assume that the aircraft is constant at 30°C. Find the air speed of the aircraft. A headwind blows at 10 m/s. How long after the aircraft passes directly overhead does its sound reach a point on the ground ?

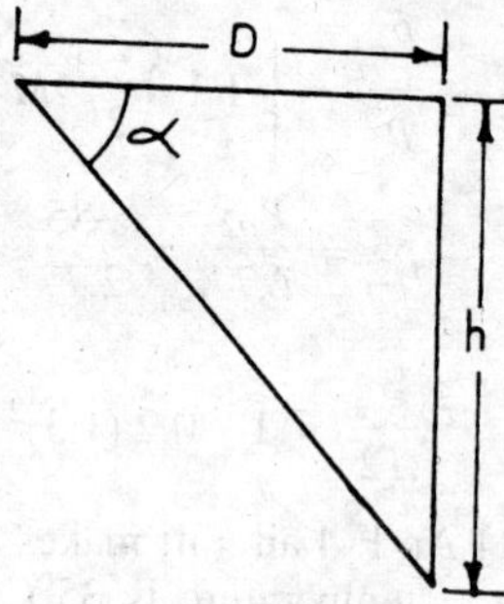

Fig 2.6. Triangle P 2.25

Solution

$$a = \sqrt{kRT}$$

$$= \sqrt{1.4 \times 287 \times 303}$$

$$= 349 \text{ m/s}$$

$$V = Ma = 1.35 \times 349 = 471 \ m/s$$ Ans

The airspeed is the velocity of the plane relative to the air.

The ground speed is then $V_p = 471 - 10 = 461$ m/s.

From the instant the aircraft is directly overhead until the Mach cane reaches the ground, the plane travels a distance, D, at speed $V_p = 461$ m/s.

Therefore, time $t = D/V_p$ and $D = h/\tan \propto$

Where $\propto = \sin^{-1}(1/M)$

$$\propto = \sin^{-1}(1/1.35) = 47.8^o$$

then, $$t = D/V_p = \frac{h}{V_p \tan\propto} = \frac{3000}{461 \tan 47.8} = 5.9 \ s$$ Ans

2.26 The concorde supersonic transport cruises at M = 2.2 at an altitude of 17 km on a standard day. How long after the aircraft passes directly above a ground observer will it be before the sound of the aircraft is heard ?

Solution

At an altitude of 17 km, t = – 56.3°C

$\therefore$ T = 273 – 56.3 = 216.7 K

$$a = \sqrt{kRT} = \sqrt{1.4 \times 287 + 216.7} = 295 \text{ m/s}$$

$$V = Ma = 2.2 \times 295 = 649 \ m/s$$

$$\propto = \sin^{-1}\left(\frac{1}{M}\right) = \sin^{-1}\left(\frac{1}{2.2}\right) = 27^o$$

See Fig 2.6

$$t = \frac{D}{V} = \frac{h}{V \tan\alpha} = \frac{17000}{649 \tan 27^o}$$

$= 51.4$ s Ans

Note :— However, the speed of sound varies over the altitude because the temperature varies with altitude.

EXERCISES

A. THEORY

2.1 Define the following terms.

Mach number
Mach angle
Mach line
Mach cone
Subsonic flow
Sonic flow
Supersonic flow

2.2 Derive an expression for speed of sound for an adiabatic process.

2.3 What is silence zone during the disturbance which propogates when an object moves in still air ? When does it occur ?

2.4 Describe Mach angle, Mach cone and Mach line with the help of sketches.

2.5 What do you mean by compressibility correction factor (ccf). Find an expression.

2.6 Show by means of diagrams the nature of propogation of disturbances in compressible flow when M < 1, M = 1 and M > 1.

2.7 Explain Mach cone

2.8 Derive an expression for speed of sound in a fluid medium confined in a rigid straight pipe of uniform cross section.

2.9 Explain what are meant by

(a) Pitot static compressibility correction in a compressible fluid flow.

(b) the local isentropic stagnation properties.

2.10 Define and state the significance of Mach number in fluid flow. Up to what limit of Mach number can we neglect the compressibility effects and treat the fluid as incompressible.

2.11 A projectile travels in a stationary air with a steady velocity. Determine and draw the nature of propogation of disturbance which this profile will made when M < 1, M = 1 and M > 1.

2.12 State the area velocity relationship for one dimensional. Show the effects of variation of area for subsonic, supersonic and sonic flows.

2.13 What is Mach number ? How is this parameter used in classification of compressible flow regimes.

2.14 What do you understand by subsonic, sonic and supersonic flows ?

2.15 Under what conditions can be air be treated as an incompressible fluid ?

2.16 Show that the velocity in a compressible fluid is given by a = $\sqrt{k/\rho}$ where k is volume modulus of elasticity of fluid.

2.17 For frictionless adiabatic flow, show that the stagnation pressure at a given point is given by

$$\frac{p_o}{p_a} = 1 + \frac{M^2}{4} + \frac{2-k}{24} M^4 + \text{------}$$

Where p_o and p_a are stagnation and ambient pressure respectively and

$$M = \frac{V}{a} = \frac{V}{\sqrt{\frac{k}{\rho}}}$$

2.18 Under what circumstances the compressibility correction factor can not be used. Explain.

2.19 Draw the sketches of Nozzle and Diffusers for subsonic and sonic conditions.

2.20 Explain, why the flow is considered incompressible for low Mach number. Give the values.

2.21 Show that for the acoustic velocity to be expressed in m/s,

$$a \approx 20.1 \sqrt{T}$$

Assume air to be a perfect gas.

2.22 Find the value of the constant C in the equation,

$$a = C\sqrt{T}$$

in order that the acoustic velocity be given in m/s for air.

2.23 Derive an expression for acoustic velocity

$$a = \sqrt{kRT}$$

The fluid has elastic properties.

2.24 Show that for sonic flow the deviation between the compressible and incompressible flow values of the pressure coefficient of a perfect gas (k = 1.4) is about 27.5 %.

2.25 We have derived equations for the ratio, p_o/p, for both compressible and incompressible flow. By writing both equations in terms of Mach number, compare their behaviour. Find the Mach number below which the two equations agree with engineering accuracy.

2.26 Show that for air $p^*/p_o^* = 0.5283$.

2.27 For a perfect gas, show that the relative pressure change across compression pulse is

$$\frac{dp}{p} = k\frac{dV}{a}$$

and the relative temperature change is

$$\frac{dT}{T} = (k-1)\frac{dV}{a}$$

2.28 Derive for isentropic flow.

(a) $$\frac{T_o}{T} = 1 + \frac{k-1}{2}M^2$$

(b) $$\frac{p_o}{p} = \left(1 + \frac{k-1}{2}M^2\right)^{k/k-1}$$

and (c) $$\frac{\rho_o}{\rho} = \left(1 + \frac{k-1}{2}M^2\right)^{1/k-1}$$

2.29 Derive for isentropic flow

$$p_o = p + \frac{\rho V^2}{2}\left(1 + \frac{M^2}{2} + \frac{2-k}{24}M^4 + \text{------}\right)$$

2.30 Derive for one dimensional compressible flow

$$\frac{dM}{M} = \frac{dV}{V} - \frac{1}{2}\left(\frac{dT}{T}\right)$$

2.31 Derive for isentropic flow

(a) $$\frac{dT}{T} = (1-k)M^2\frac{dV}{V}$$

(b) $$\frac{kR}{k-1}dT + V\,dV = 0$$

(c) $$\frac{a^2}{k-1} + \frac{V^2}{2} = \frac{a_o^2}{k-1}$$

(d) $$V_{max}^2 = \frac{2}{k-1}a_o^2$$

(e) $a^{*2} = \frac{2}{k+1} a_o^2$

(f) $\frac{kR}{k-1} T + \frac{V^2}{2} = \frac{kR}{k-1} T_o$

2.32 Show that for isentropic flow

$$V_{max} = \sqrt{\frac{2a^2}{k-1} + V^2}$$

2.33 Show that for isentropic flow

$$a^* = \sqrt{\frac{2a^2 + V^2 (k-1)}{k+1}}$$

2.34 Show that the maximum velocity that can result from the flow of a perfect gas from a reservoir is

$$V_{max} = \sqrt{\frac{2}{k-1}}\, a_o$$

Where a_o is the acoustic velocity in the gas at reservoir conditions. Is the corresponding Mach number $\sqrt{\frac{2}{k-1}}$? Explain.

2.35 Show that for isentropic flow of a perfect gas, the pressure, temperature, and density, when made dimensionless with respect to the corresponding critical values, are given by

$$\frac{p}{p^*} = \left[\frac{k+1}{2\left(1 + \frac{k-1}{2} M^2\right)}\right]^{k/k-1}$$

$$\frac{T}{T^*} = \frac{k+1}{2\left(1 + \frac{k-1}{2} M^2\right)}$$

$$\frac{\rho}{\rho^*} = \left[\frac{k+1}{2\left(1 + \frac{k-1}{2} M^2\right)}\right]^{1/k-1}$$

2.36 Establish the relation

$$M = \frac{V}{V_{max}} \sqrt{\frac{2/(k-1)}{1 - V^2/V_{max}^2}}$$

2.37 Derive for one dimensional isentropic flow

$$\frac{dA}{A} = \frac{dp}{\rho V^2}\left(1 - M^2\right)$$

2.38 For an ideal gas prove that

$$\frac{T}{T_o} = 1 - \frac{V^2}{V^2_{max}}$$

From this show that T/T_o is indepndent of the thermodynamics properties of gas itself.

2.39 For an adiabatic flow of a perfect gas, show that

(a) $$\frac{T}{T_o} = 1 - \frac{k-1}{k+1} M^{*2}$$

(b) $$\frac{a}{a_o} = \frac{M^2}{1 + \frac{k-1}{2} M^2}$$

2.40 Show that

$$M^{*2} = \frac{\frac{M^2}{2}(k+1)}{M^2 [(k-1)/2] + 1}$$

2.41 Show that

$$M = \frac{M^* \sqrt{2/k+1}}{\sqrt{1 - \frac{k-1}{k+1} M^{*2}}}$$

B. UNSOLVED PROBLEMS

2.1 Find the Mach number of a rocket travelling in standard air with a speed of 1600 km/hr. [1.306]

2.2 Find the velocity of a bullet fired in standard air if its Mach angle is 30°. [680.4 m/s]

2.3 Consider a missile travelling at a 45 km altitude with a Mach number of 7. Find the stagnation temperature at the nose of the missile. Assume the temperature at 45 km altitude is 45^oC [3120 K].

2.4 Air at a temperature of 300 K is flowing at a Mach number of 2. Find the velocity of flow. What will be the Mach angle ? [696 m/s, 30°]

2.5 A jet engine travels at 1100 km/hr through still air having a pressure of 0.7 bar and temperature of – 5^oC. Assume wind velocity is negligible. Find the

Mach number. What will be the pressure, temperature and density of air at the stagnation point on the nose of the jet. [M = 0.931, T_o = 314.45 K, P_o = 1.22 bar; ρ_o = 1.3572 kg/m^3]

2.6 The velocity of sound in a gas is found experimentally to be 600 m/s When the gas pressure is 1.75 bar and the density is 0.682 kg/m^3. What is the value of the specific heat ratio k for this gas ? [1.4].

2.7 An aircraft flies at a speed of 960 km/hr through air at a pressure of 82 kPa and temperature of 0°C. Calculate the Mach number of the craft. [M = 0.806]

2.8 Air at a temperature of 20°C is flowing with a speed of 150 m/s. A bullet is fired into the airstream with a speed of 800 m/s. (The direction of the bullet is opposite to that of the air). Calculate :

(a) The Mach number of the air flow

(b) The Mach number of the bullet if it were fired in still air.

(c) The Mach number of the bullet with respect to the moving air. [M_{air} = 0.437, M_{bullet} = 2.33, M_{bullet} = 2.77]

2.9 An airplane flies at a speed of 180 m/s at an altitude of 500 m, where the temperature of 20°C. The plane climbs to 15 km, where the temperature is – 56°C and levels off at a speed of 320 m/s. Calculate the Mach number of flight in both cases. [M_1 = 0.525, M_2 = 1.08].

2.10 A photograph of a bullet shows a Mach angle of 28°. Determine the speed of the bullet for standard air. [725 m/s]

2.11 Air at 25°C is flowing at a Mach number of 1.9. Determine the air speed and the Mach angle. [657 m/s, $\propto$ = 31.8°].

2.12 An aeroplane is in steady flight at a Mach number of 0.95. If the air temperature is – 10°C. Calculate the temperature at the stagnation point on the leading edge of the wing and also at a point just outside the boundary layer where the local Mach number is 1.25. [T_o = 310 K, 236 K].

2.13 A pitot static tube is used to measure the velocity of an air stream. Determine the velocity of an air stream. Determine the numerical factor by which the actual pressure difference shown by the gauge must be multiplied to allow for Compressibility when the value of Mach number is 0.8. [CCF = 1.16 approx.]

2.14 A rocket travels in air of pressure 1.033 bar at 15°C at a velocity of 1650 km/hr. Find the Mach Number and the Mach angle. [M = 1.347, $\propto$ = 47° 56′].

2.15 A jet propelled aircraft is flying at 1100 km/hr at sea level. Find the Mach number if the temperature of atmosphere is 21°C. [M = 0.808].

2.16 In a test section of a supersonic tunnel a pitot static tube indicates a static pressure of 0.8 bar while the difference between the static and stagnation pressure was 110 mm of mercury. Find the Mach number and the velocity of air. [0.324, 130 m/s].

2.17 A DC-10 air craft cruises at an altitude of 12 km on a standard day. A pitot static tube located on the nose of the aircraft measures the stagnation and static pressures as 29.6 kPa and 19.4 kPa, respectively. Calculate (a) the flight Mach number of the aircraft, (b) the speed of the aircraft, and (c) the stagnation temperature that would be sensed by a probe on the aircraft. [M = 0.801, V = 236 m/s, T_o = 245 K].

2.18 The Anglo–French "Concorde" supersonic transport cruises at M = 2.2 at an altitude of 20 km. Evaluate the speed of sound, aircraft flight speed, and Mach angle for these conditions. Compare the aircraft speed of to the muzzle speed of a 0.22 caliber rifle bullet (460 m/s). What is the maximum air temperature at stagnation points on the aircraft structure ? [a = 295 m/s, V = 649 m/s, $\propto$ = 27°, V/V_b = 1.41, T_o = 426 K].

2.19 A transonic wind tunnel operates at test section Mach number, M = 1.2. Operating conditions in the test section are –34°C and 41.8 kPa. Calculate the local isentropic stagnation conditions for this flow. A normal shock forms ahead of a pitot static tube placed in the flow stream. The total-head tube measures a stagnation pressure of 100.6 kPa. Determine the loss in stagnation pressure across the normal shock. By what percentage is the local isentropic stagnation pressure reduced compared to the upstream value ? [T_o = 308 K, p_{o1} = 101.3635 kPa, $p_{o1} - p_{o2}$ = 0.8 kPa % = 0.79]

2.20 Consider the steady adiabatic flow of air through a long straight pipe with cross sectional area 0.05 m^2. At the inlet section the air is at an absolute pressure of 200 kPa, 60°C, and velocity of 146 m/s. At a section downstream the air is at 95.6 kPa and has a velocity of 280 m/s. Determine $p_{o_1}, p_{o_2}, T_{o_1}, T_{o_2}$. [223 kPa, 344 k, 145 kPa, $T_{o1} = T_{o2}$ = 344 K].

2.21 Air enters a combustion chamber at M_1 = 0.2 T_1 = 580 K, and P_1 = 1 MPa. Heat addition in the combustion chamber causes the fluid properties at the exit to be M_2 = 0.4, T_2 = 1727 K and P_2 = 862.7 kPa Determine $p_{o_1}, p_{o_2}, T_{o_1}, T_{o_2}$. [10.28 bar, 9.632 bar, 584.6 K, 1782 K]

2.22 The tires on a modern high weight bicycle are inflated to 800 kPa. Consider a day when a tire is in equilibrium with the ambient at 37°C. Calculate the critical conditions that correspond to these stagnation conditions. [T^* = 253 K, p^* = 476 kPa, V^* = 319 m/s].

2.23 Stagnation conditions in solid propellant rocket motor are T_o = 3500 K and p_o = 40 MPa. Critical conditions occur in the throat of the rocket nozzle where the Mach number is equal to one. Evaluate the pressure, temperature, and flow speed at the throat of the rocket nozzle. Take R = 323 J/kg K, k = 1.2 for the exhaust gases. [T_o = 3180 K, p_o = 22.6 MPa, V_o = 1110 m/s].

3

VARIABLE AREA FLOW

3.1 Introduction

In this section we consider steady inviscid isentropic quasi one dimensional, one directinal compressible flow through a channel of variable cross sectional area. Variable area sections may be convergent, divergent, or convergent divergent. The flow patterns through such sections may be such as to produce either accelerations or decelerations of the flowing fluid. If the flow is accelerated, the section is called a nozzle, or if it is decelerated, it is called a diffuser. Whether a certain flow section will act as a nozzle or as a diffuser depends not only on its contour but also on the entering character of the fluid. We have already developed all the basic relations needed for analyzing this flow.

3.1.1 Effect of area variation on compressible flow

When fluid flows steadily through a conduit that has a flow cross section area that varies with axial distance, the conservation of mass (continuity) equation

$$\dot{m} = \rho A V = constant$$

Taking the natural logarithm of both sides, yields

$$\ln \rho + \ln A + \ln V = \ln constant$$

Differentiating, we get

$$\frac{d\rho}{\rho} + \frac{dA}{A} + \frac{dV}{V} = 0$$

Solving the above equation for $\frac{dA}{A}$ gives

$$\frac{dA}{A} = -\frac{dV}{V} - \frac{d\rho}{\rho} \quad (3.1)$$

The Euler's equation of motion for compressible fluid is given by

$$\frac{dp}{\rho} + V\,dV + g\,dz = 0$$

Neglecting gravitational forces

$$\frac{dp}{\rho} + V\,dV = 0$$

or, $dp = -\rho\, V\, dV$

Dividing by $\rho\, V^2$, we obtain

$$\frac{dp}{\rho\, V^2} = -\frac{dV}{V} \tag{3.2}$$

Substituting into equation (3.1)

$$\frac{dA}{A} = \frac{dp}{\rho V^2} - \frac{d\rho}{\rho}$$

$$= \frac{dp}{\rho V^2}\left[1 - \frac{V^2}{dp/d\rho}\right]$$

From equation (2.3)

$$a^2 = \frac{dp}{d\rho}$$

$$= \frac{dp}{\rho\, V^2}\left[1 - \frac{V^2}{a^2}\right]$$

$$= \frac{dp}{\rho\, V^2}[1 - M^2] \tag{3.3}$$

From equation (3.3) we see that for $M < 1$ an area change causes a pressure change of the same sign (positive dA means positive dp for $M < 1$) for $M > 1$ an area change causes a pressure change of oppposite sign.

We know from equation (2.3) $a^2 = \frac{dp}{d\rho}$, therefore, $dp = a^2\, d\rho$.

Substituting in equation (18.64)

$$\frac{dA}{A} = \frac{a^2 d\rho}{\rho V^2}(1 - M^2)$$

or, $$\frac{d\rho}{\rho} = \frac{dA}{A}\left(\frac{V^2}{a^2}\right)\left(\frac{1}{1 - M^2}\right)$$

$$= \frac{dA}{A}\left(\frac{M^2}{1 - M^2}\right) \tag{3.4}$$

Coming equation (3.2) and (3.3)

$$\frac{dA}{A} = -\frac{dV}{V}(1 - M^2)$$

$$\frac{dA}{A} = -\frac{dV}{V}(M^2 - 1) \tag{3.5}$$

Equation (3.5) is known as the Hugoniot equation and significant equation for one dimensional, isentropic flow.

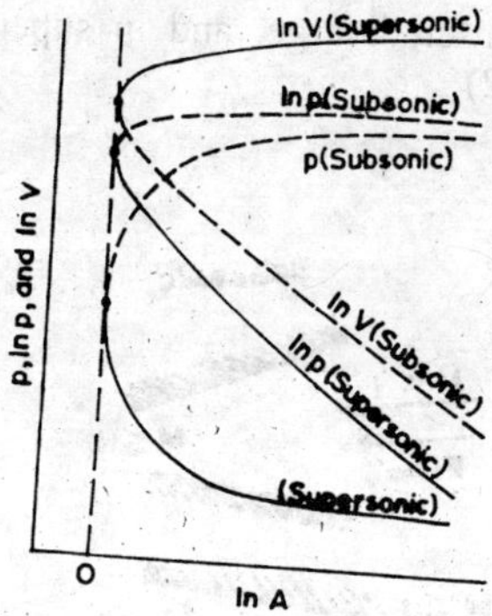

Fig 3.1 Pressure, density and velocity variaiton in isentropic flow.

Fig (3.1) shows the pressures, density and velocity in isentropic flow.

For subsonic flow, (M < 1), Fig 3.1 shows

(a) The pressure increases with an area increase.

(b) The density increases with an area increase.

(c) The velocity decreases with an area increase.

For supersonic flow, (M > 1), Fig 3.1 shows

(a) The pressure decreases with an area increase.

(b) The density decreases with an area increase.

(c) The velocity increases with an area increase.

For flow at sonic speeds, (M = 1)

(a) $\frac{dA}{dp} = 0 = \frac{dA}{d\rho} = \frac{dA}{dV}$; thus the area may be either a maximum or a minimum. Fig 3.1 shows that at M = 1, it must be a minimum.

(b) As the sonic condition is approached (from either side) the pressure changes (and also density changes) becomes very large for even a small area change.

(c) Combining equations (3.4) and (3.5),

$$\frac{d\rho}{\rho} = -M^2\frac{dV}{V} \tag{3.6}$$

Thus near sonic velocities the change in density and the change in velocity will compensate.

A nozzle is a device in which we want to gain a velocity increase at the expense of pressure (pressure decrease). To accomplish this, in subsonic flow the nozzle area must decrease, and in supersonic flow the area must increase.

A diffuser is a device in which we want to gain a pressure increase at the expense of velocity (velocity decrease). This may be accomplished in subsonic flow by making the diffuser diverge, and in supersonic flow by making the diffuser converge. (Fig 3.2)

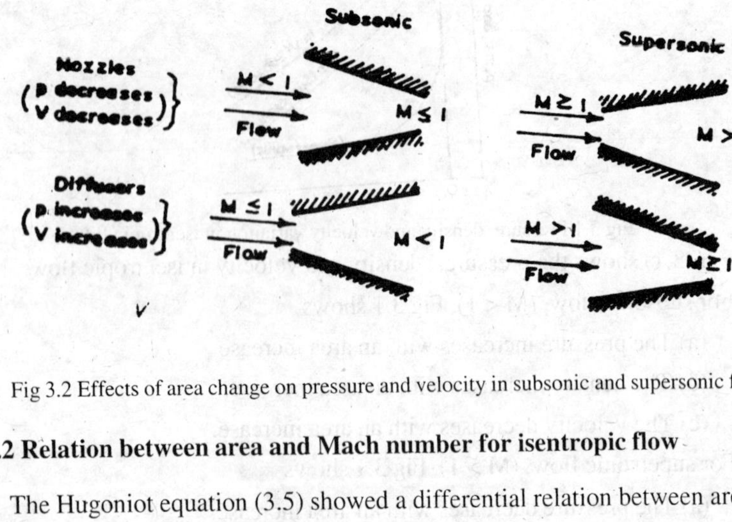

Fig 3.2 Effects of area change on pressure and velocity in subsonic and supersonic flow.

3.1.2 Relation between area and Mach number for isentropic flow

The Hugoniot equation (3.5) showed a differential relation between area and velocity.

Since a relation is sought between area and Mach number and since the Mach number is a dimensional quantity, the area must be rendered dimensionless. This may be done by referring the flow area to some constant area that is characteristics to the flow. It is necessary for a passage to have a section of minimum area (a throat) to accelerate a flow isentropically from rest to a Mach number greater than unity. Furthermore, in such a flow the Mach number is unity at the throat. If the area at which the Mach number is unity is designated by A^*, then it is possible to express the contour of a passage in terms of the area ratio A/A^*.

The continuity equation can be written as

$$\rho A V = constant = \rho^* A^* V^*$$

then

$$\frac{A}{A^*} = \frac{\rho^*}{\rho}\frac{V^*}{V} = \frac{\rho^*}{\rho}\frac{C^*}{Ma} = \frac{1}{M}\frac{\rho^*}{\rho}\sqrt{\frac{T^*}{T}}$$

$$\frac{A}{A^*} = \frac{1}{M}\frac{\rho^*}{\rho_o}\frac{\rho_o}{\rho}\sqrt{\frac{T^*/T_o}{T/T_o}}$$

$$= \frac{1}{M}\left[\frac{\left[1+\frac{k-1}{2}M^2\right]^{1/k-1}}{\left[1+\frac{k-1}{2}\right]^{1/k-1}}\right]\left[\frac{1+\frac{k-1}{2}M^2}{1+\frac{k-1}{2}}\right]^{1/2}$$

$$= \frac{1}{M}\left[\frac{1+\frac{k-1}{2}M^2}{\frac{k+1}{2}}\right]^{1/k-1}\left[\frac{1+\frac{k-1}{2}M^2}{\frac{k+1}{2}}\right]^{1/2}$$

$$\frac{A}{A^*} = \frac{1}{M}\left[\frac{1+\frac{k-1}{2}M^2}{\frac{k+1}{2}}\right]^{\frac{k+1}{2(k-1)}} \qquad (3.7\text{ (a)})$$

$$= \frac{1}{M}\left[\left(\frac{2}{k+1}\right)\left(1+\frac{k-1}{2}M^2\right)\right]^{\frac{k+1}{2(k-1)}} \qquad (3.7\text{ (b)})$$

$$= \frac{1}{M}\left[\frac{2}{k+1}+\frac{k-1}{k+1}M^2\right]^{\frac{k+1}{2(k-1)}} \qquad (3.7\text{ (c)})$$

From equation (3.7) we see that a choice of M gives a unique value of A/A^* Fig 3.3. Note that the curve is double valued; for a given value of A/A^* other than unity, there are two possible values of Mach number. To each M a single value of A/A^* exists whereas each A/A^* corresponds to two values of M.

As in the case of the stagnation and critical states discussed earlier, one may use euqation (3.7) to define a local A^* associated with each point in a flow where both M and A are known. This local A^* would be the flow area that would be necessary to accelerate or decelerate the flow at that point to M = 1 by means of an isentropic process. The area A^* need not actually exist, but its definition as a local quantity may be useful. In the event that th flow is isentropic, the local A^* associated with each point will be the same. Thus M and A are known at one point in an isentropic flow, A^* may be ascertained and equation (3.7) used to find M as a function of A elsewhere in the flow.

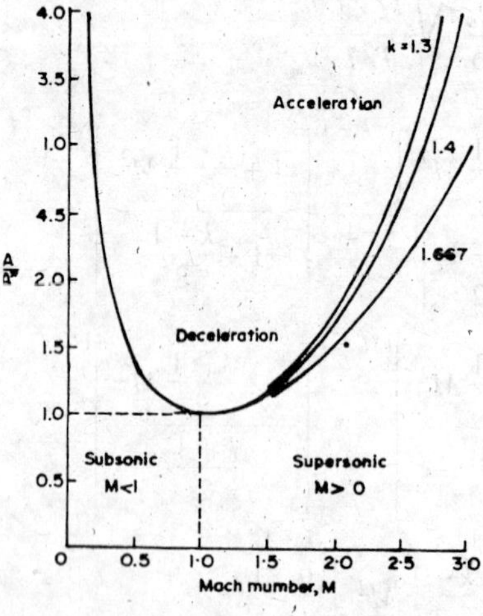

Fig 3.3 Area ratio A/A^* as a function of Mach number and k.

In general the ratio of the areas at two different points in a flow may be expressed by

$$\frac{A_2}{A_1} = \frac{A_2}{A_2^*} \times \frac{A_1^*}{A_1} \times \frac{A_2^*}{A_1^*} \tag{3.8}$$

The first two terms on the right side of equation (3.8) are depedent only on the Mach number at the two points in question. The last term of the equation depends on the degree of departure of the flow from isentropy between the points in question. For isentropic flow $A_2^*/A_1^* = 1$, and the area ratio is dependent on the Mach number only

$$\frac{A_2}{A_1} = \frac{A_2}{A_2^*} \times \frac{A_1^*}{A_1}$$

In many compressible flow calculations the function $\frac{A}{A^*}\frac{p}{p_o}$ occurs frequently. This can also be obtained as a function of Mach number from equation (2.16) and (3.7).

$$\frac{A}{A^*}\frac{p}{p_o} = \frac{1}{M}\left(\frac{2}{k+1}+\frac{k-1}{k+1}M^2\right)^{\frac{k+1}{2(k-1)}}\frac{p}{p_o}$$

$$= \frac{\frac{1}{M}\sqrt{\left(\frac{2}{k+1}+\frac{k-1}{k+1}M^2\right)^{(k+1)/(k-1)}}}{\left(1+\frac{k-1}{2}M^2\right)^{k/k-1}}$$

This on simplification and rearrangement yields

$$\frac{A}{A^*}\frac{p}{p_o} = \frac{1}{M}\frac{\sqrt{\left(\frac{2}{k+1}\right)^{k+1/(k-1)}}}{\left(1+\frac{k-1}{2}M^2\right)} \tag{3.9}$$

3.1.3 Mass flow rate as a function ratio or Mach number

We now turn our attention to a consideration of the mass flow rate per unit area, m/A. From the continuity equation we proceed as follows

$$\frac{\dot{m}}{A} = \rho V = \frac{pV}{RT} = \frac{pV}{RT}\sqrt{\frac{kT_o}{kT_o}}$$

$$= \frac{pV}{\sqrt{kRT}}\sqrt{\frac{k}{R}}\sqrt{\frac{T_o}{T}}\sqrt{\frac{1}{T_o}}$$

$$= \frac{pM}{\sqrt{T_o}}\sqrt{\frac{k}{R}}\sqrt{1+\frac{k-1}{2}M^2} \tag{3.10}$$

$$= \frac{p}{p_o}\frac{p_o M}{\sqrt{T_o}}\sqrt{\frac{k}{R}}\sqrt{1+\frac{k-1}{2}M^2}$$

$$= \frac{p_o}{\sqrt{T_o}}M\sqrt{\frac{k}{R}}\left(1+\frac{k-1}{2}M^2\right)^{-k/k-1}\sqrt{\left(1+\frac{k-1}{2}M^2\right)}$$

$$= \frac{p_o M}{\sqrt{T_o}}\sqrt{\frac{k}{R}}\left[\left(1+\frac{k-1}{2}M^2\right)\right]^{-\frac{(k+1)}{2(k-1)}}$$

$$= \frac{p_o}{\sqrt{T_o}}\sqrt{k/R}\frac{M}{\left(1+\frac{k-1}{2}M^2\right)^{\frac{(k+1)}{2(k-1)}}} \tag{3.11}$$

Since the mass flow rate passes through a maximum value when M = 1, the maximum mass flow rate for a given p_o and To is passed through a given area when M = 1. The maximum mass flow rate can be found by setting M = 1 in equation (3.11)

$$\dot{m}_{max} = A^* \frac{p_o}{\sqrt{T_o}}\sqrt{\frac{k}{R}} \frac{1}{\left(\frac{k+1}{2}\right)^{(k+1)/2(k-1)}} \qquad (3.12)$$

It is to be noted that A has been replaced in equation (3.12) by A* since M = 1 can occur only at a throat. Putting k = 1.4, R = 287 J/kg K the non dimensional mass flow rate is obtained.

$$\frac{\dot{m}_{max}\sqrt{T_o}}{p_o A^*} = \sqrt{\frac{k}{R}}\left(\frac{2}{k+1}\right)^{(k+1)/2(k-1)}$$

$$= \sqrt{\frac{1.4}{287}}\left(\frac{2}{2.4}\right)^{2.4/2\times 0.4}$$

$$= 0.06984 \times (0.8333)^3 = 0.06984 \times 0.5787$$

$$= 0.040418 \qquad (3.13)$$

Where $\dot{m}_{max}$ is in kg/s, T_o is in K, A^* is in m^2, p_o is in N/m^2

This is also known as Fliegner's formula; he developed this expression empiricially much earlier than its theoretical derivation. Fliegner verified this by experiments with very close agreement.

3.2 FLOW THROUGH NOZZLES

Flow through nozzle occurs in a variety of engineering applications. In steam turbine design, gas turbine design. Nozzles used as propelling device in some air breathing engines are converging as well as converging-diverging. For values of about $p_o/p \approx 3$, most aircrafts use converging nozzle as propelling device becuase the thrust produced by a converging nozzle is as large as convergent divergent nozzle and shock losses are eliminated. For values of pressure ratios p_o/p as high as 10 to 20 for flight Mach numbers in the range of 2 to 3 a canvergent-divergent nozzle is employed. Nozzles used in rockets are invariably converging diverging becuase of very large pressure ratios and very high exit velocities. The characteristics feautre of such nozzles is large area ratio of the order of 10 to 20.

Convergent nozzles are used for subsonic and sonic flows. They can also be used as flow measuring and flow regulating devices.

Convergent-divergent nozzles are used for supersonic flows. It is comparatively difficult to obtain convergent-divergent passages in compressors and turbine blade rows.

3.2.1 Convergent nozzle

Let us consider a convergent nozzle as shown in Fig 3.4. The conditions upstream are the stagnation conditions, which are assumed to be constant. The pressure at the exit plane of the nozzle is designated p_e, and the back pressure p_b.

Let us consider how the mass flow rate $\dot{m}$ and the exist plane pressure p_e / p_o vary as the back pressure p_b is decreased. These quantities are plotted in Fig 3.4.

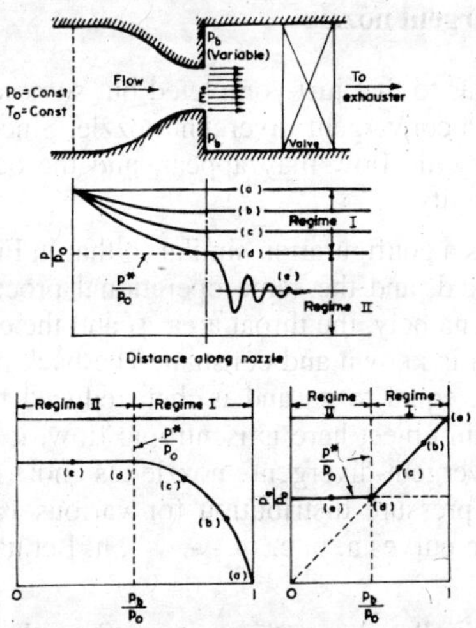

Fig 3.4 Flow through a convergent nozzle

If $p_b/p_o = 1$ (condition a, Fig 3.4) then the pressure is constant throughout the nozzle and there is no flow.

If p_b is slightly reduced from this value (condition b, Fig 3.4) a subsonic flow will be established with the exit pressure p_e equal to the back pressure p_b.

If p_b is reduced further (condition c, Fig 3.4), the flow remains subsonic with $p_b = p_e$, but the mass rate increases.

The mass flow rate increase continues until p_b/p_o reaches the critical pressure ratio $\left(\frac{p_b}{p} = \frac{p^*}{p_o} = \frac{p_e}{p_o} \right)$ where $M_e = 1$ (condition d Fig 3.4)

If pb reduced further, (condition d, Fig 3.4), the pressure p_e can not be become less than p^* since M_e stays at 1. Therefore the mass flow rate remains constant and the pressure distribution inside the nozzle remains the same as for condition d.

The pressure distribution in the chamber outside the nozzle p_b/p_e can not be predicted accurately by one dimensional model and is indicated by a wavy line.

The entire flow is divided into two regimes. The regime I includes the cases between a to c where the flow is subsonic $M_e < 1$ and the mass flow rate is dependent on pressure ratio p_b/p_o and regime II includes cases d and e with Mach number at exit Me = 1 and the mass flow rate is independent of p_b/p_o but depends upon the stagnation conditions.

3.2.2 Convergent-divergent nozzle

A discussion similar to that just completed but somewhat less detailed will now be carried out for a convergent-divergent nozzle. Since a relative minimum flow area exists, supersonic flow may appear, and the occurance of a normal shock becomes a possibility.

Fig 3.5 illustrations a configuration similar to that in Fig 3.4. The same flow characteristics are plotted, and the same operational procedure is imagined. A fixed nozzle geometry, namely, the throat area A_t and the exit area A_e are given. The supply (p_o and T_o) is known and constant. The back pressure p_b, presumed specified, is originally, equal to p_o and is then reduced to successively lower values. Though the main object here is isentropic flow, it will be observed that flow through a convergent-divergent nozzle is not isentropic under all conditions. The static pressure distribution for various values of the pressure ratio p_b/p_o are shown in curves a, b, c, - - - - - - -, h. Let us consider each of the cases shown.

With the valve initially closed there is no flow through the nozzle, the pressure is constant at p_b. Opening the valve slightly (p_b slightly less than p_o) produces the pressure distribution cruve a. If the flow rate is low enough, at all points on this curve the flow will be subsonic and essentially incompressible. Under these conditions, the convergent divergent nozzle will behave like a venturi, with flow accelerating in the converging portion until a point of maximum velocity and minimum pressure is reached at the throat, then decelerating in the diverging portion to the nozzle exit. If p_b is lowered to condition b, the mass flow rate increases but the flow pattern does not change.

If p_b is lowered to the value corresponding to condition c, flow at the throat becomes sonic but again it diffuses to the back pressure $p_b = p_e$. Condition g corresponds to the flow continuously accelerating through the nozzle. In this case the pressure ratio p_e/p_o is called the design pressure ratio of the nozzle. The mass flow rate is the same for conditions c and g and corresponds to the maximum flow rate since M = 1 at throat for both cases.

The flow corresponding to the back pressure between conditions c and g of Fig 3.5 involves shock discontinuities.

As p_b is reduced below the corresponding to condition C, the flow will accelerate supersonically for a short distance after the throat and a shock will appear in the divergent passage. This shock decelerates the flow to the higher pressure and the flow then decelerates isentropically to the exit pressure, $p_b = p_e$.

As p_b is further reduced the shock moves down the nozzle until p_b = pressure at case e, the limiting case of regime II, where the shock appears remove at the exit from the nozzle and the flow is every where supersonic in the divergent part. In regime $p_e = p_b$ and mass flow rate is every where constant and equal to the maximum mass flow rate through the passage.

As regime III is entered by reducing p_b below the pressure at case e, the convergent divergent nozzle exit pressure stays at g and the flow is everywhere supersonic in the divergent nozzle. The convergent divergent nozzle.

For condition g there are no shocks anywhere in the flow and $p_b = p_e$. The exit pressure for this case is the design exit pressure of the convergent divergent nozzle. When p_b drops below this value (regime IV), oblique expansions waves are formed in the reservoir around the exit of the convergent divergent nozzle.

In both regimes III and IV the flow pattern in the convergent divergent nozzle is independent of the back pressure p_b and is the same as for the design conditions.

The convergent divergent nozzle is chocked in regimes II, III and IV and the mass flow rate, $\dot{m}$, is independent of the back pressure and is maximum. Only in regime I can the mass flow rate be changed by variation in the back pressure.

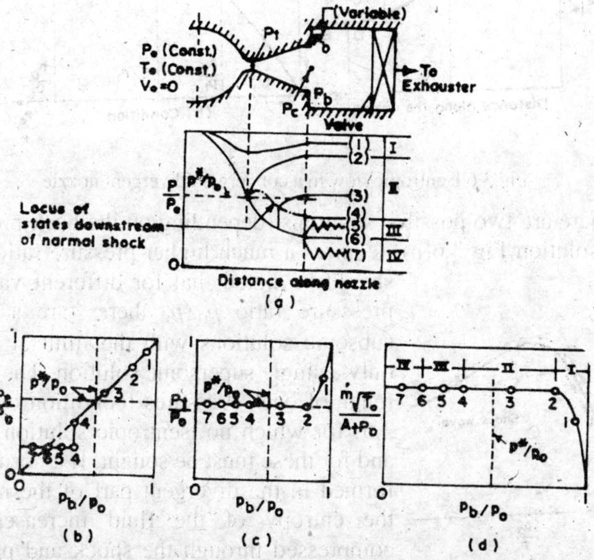

Fig 3.5 Flow through a convergent divergent duct

In region where flow can be assumed to be isentropic, the pressure distribution can be found from $\left[\frac{\dot{m}\sqrt{T_o}}{A\,p_o}\right]_{max}$

or from Fig 3.6 which shows the variation of mass flow function $\frac{\dot{m}\sqrt{T_o}}{A\,p_o}$ with p_o/p. The mass flow function increases from O at $p_o/p = 1$, reaches a maximum at M = 1 curve of $\frac{\dot{m}}{A}$ against p_o/p.

The solution for subsonic flow is shown in Fig 3.6(a). However, if M = 1 at

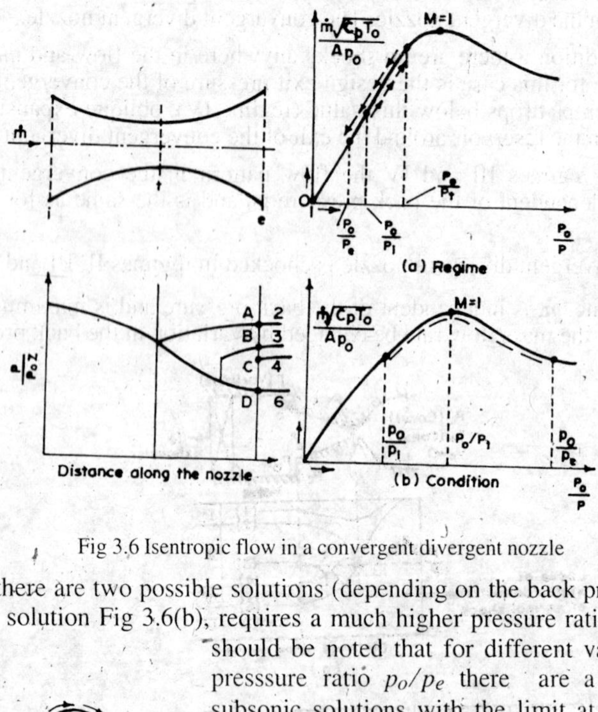

Fig 3.6 Isentropic flow in a convergent divergent nozzle

the throat there are two possible solutions (depending on the back pressure), the supersonic solution Fig 3.6(b), requires a much higher pressure ratio (P_o/P_e). It should be noted that for different values of the presssure ratio p_o/p_e there are a number of subsonic solutions with the limit at M = 1 but only a single supersonic solution. There is a wide range of back pressures (condition c to g of Fig 3.5) for which no isentropic solution is possible and for these must be sought. If a normal shock is formed in the divergent part of the nozzle, then the entropy of the fluid increases as it is compressed through the shock and p_o decreases

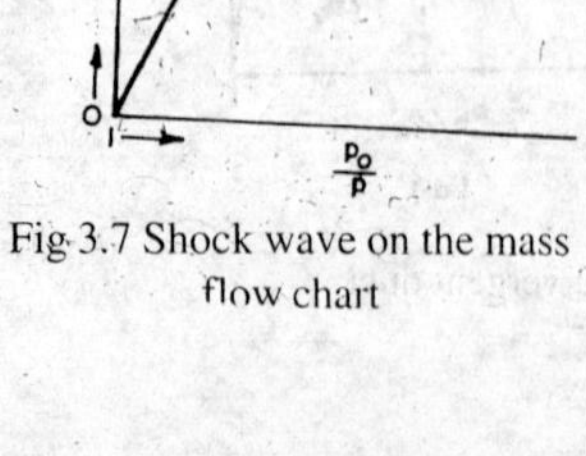

Fig 3.7 Shock wave on the mass flow chart

on the graphs $\frac{\dot{m}\ \sqrt{T_o}}{A\ p_o}$ will increase from supersonic to a subsonic value (Fig 3.7).

In this case the gas will have a higher pressure a the exit plane than for condition g, (condition D, Fig 3.6). Allowance for the normal shock gives solutions between conditions gives solutions between conditions C and e; A weak shock near the throat where the Mach number is just greater than 1 will require a back pressure slightly lower than that for condition C. A strong shock standing at the nozzle exist, where the Mach number is maximum will give a much greater loss of stagnation pressure and will require a much smaller back pressure than for condition C.

3.3 FLOW THROUGH DIFFUSERS

The term "diffuser" is used to denote a flow passage intended to decelerate a flow, thereby producing a rise in the fluid pressure. Many instances of engineering significance exist in which it is desired to accomplish diffusion of a gas stream. Devices such as centrifugal fans and compressors accomplish their purpose by imparting kinetic energy to the gas by accelerating it with an impeller. The moving gas stream must then be diffused in order to convert this added energy into a pressure increase. Similarly, the efficiency characteristics of a compressor supplied with a high velocity gas stream requires that the gas first be decelerated by means of a diffuser. Air breathing propulsion devices for aircraft or missiles must induct the needed air from the moving atmosphere (reletively speaking) around it. This inducted stream of moving air must then be decelerated by a diffuser before the subsequent compression or combusion processes of the propulsion devices can take place.

Thus one distinguishes two basic types of diffusers : ducted diffusers, which receive their gas supply from a closed duct, and inlet diffusers, which receive their gas supply by moving through a large, free mass of gas. Such diffusers are shown in Fig 3.8 It should be noted that in the case of inlet diffusers, it is

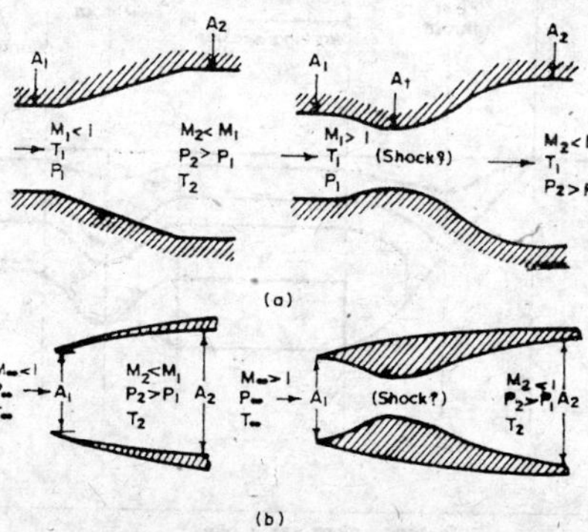

Fig 3.8 Two types of diffusers

common notation to use the subscript as to denote the properties of the ambient atmosphere through which the diffuser moves.

For Fig 3.8 (a)

3.8 (b)

3.3.1 Supersonic diffusers

When flow at the entrance to a convergent divergent passage is supersonic, then the fluid will be slowed down and compressed to a Mach number $M_t \geq 1$ at the throat. The convergent duct is then referred to as a supersonic diffuser. If the Mach number at the throat is unity the fluid may expand again to supersonic velocities, or compressed beyond the throat at subsonic velocities. Supersonic diffusers are used in wind tunnels with a supersonic working section and in jet engine intakes. We now consider the flow in these devices and for simplicity ignore all losses.

3.3.2 Wind tunnel operation with a supersonic diffusers

A simple configuration for the tunnal is shown in Fig 3.9. It consists of a convergent divergent duct, a working section, another convergent divergent duct and a compressor. When the flow is established in the tunnel, the second throat could be matched to the first throat to give nearly isentropic flow throughout the tunnel sections. However during starting the second throat must be larger than the first throat.

We will discuss in details in the Chapter 6.

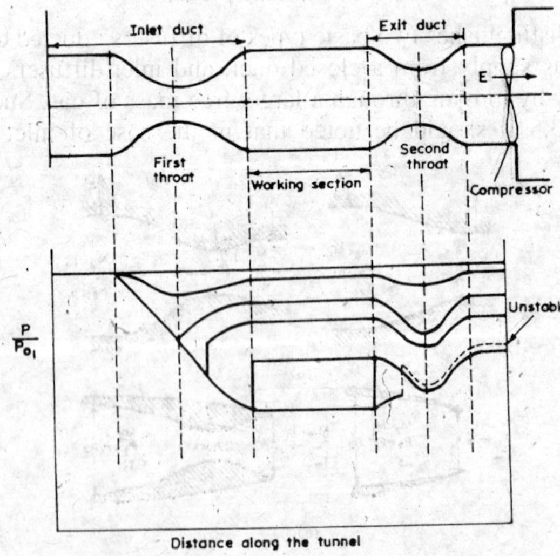

Fig 3.9 Supersonic wind tunnel

ILLUSTRATIVE EXAMPLES

3.1 A Convergent nozzle has an exit area of 500 mm^2. Air enters the nozzle with a stagnation pressure of 1000 kPa and a stagnation temperature of 360K. Determine the mass rate of flow for back pressure of 800 kPa, 528 kPa, and 300 kPa, assuming isentropic flow.

Solution The critical pressure ratio is

$$\frac{p^*}{p_o} = 0.528$$

For a back pressure 528 kPa, M = 1 at the nozzle exit and the nozzle is choked.

Decreasing the back pressure below 528 kPa will not increase the flow.

(a) For a back pressure of 528 kPa,

$$\frac{T^*}{T_o} = 0.8333 \quad \therefore \quad T^* = 300\ K$$

At the exit, $V = a = \sqrt{k\,RT}$

$= 20.05\sqrt{300} = 347.2$ m/s

$$\rho^* = \frac{p^*}{RT^*} = \frac{528}{0.287 \times 300} = 6.1324\ kg/m^3$$

At the throat,

$$\dot{m} = \rho\, A\, V$$

$$= 6.1324 \times 500 \times 10^{-6} \times 347.2 = 1.0646\ kg/s$$

(b) For a back pressure of 800 kPa,

$$\frac{p_e}{p_o} = 0.8$$

From the isentropic tables,

$$M_e = 0.573, \frac{T_e}{T_o} = 0.9381$$

$\therefore$ $T_e = 337.7$ K.

$$a_e = \sqrt{k\,RT_e}$$

$$= \sqrt{1.4 \times 287 \times 337.7} = 368.4\ m/s$$

$$V_e = M_e\, a_e = 0.573 \times 368.4 = 211.1\ m/s$$

$$\rho_e = \frac{p_e}{RT_e} = \frac{800}{0.287 \times 337.7} = 8.2542\ kg/m^3$$

At the exit section

$$\dot{m} = \rho_e A V_e$$

$$= 8.2542 \times 500 \times 10^{-6} \times 211.1 = 0.8712\ kg/s$$

(c) For a back pressure less than critical pressure, which in this case is 528 K the nozzle is choked and the mass rate is the same as that for the critical pressure. Therefore, for an exhaust pressure of 300 kPa the mass rate of flow is 1.0646 kg/s.

3.2 A Converging duct passes air steadily from standard atmospheric conditions to a receiver pipe. The throat flow cross section area of the converging duct is $1 \times 10^{-4}\ m^2$. Determine the mass flow rate through the duct if the receiver pressure is (a) 80 kPa; (b) 40 kPa. Take standard values of temperature and density are 15°C and 1.23 kg/m^3 respectively.

Solution

(a) Since the receiver pressure, is greater than the critical pressure, $p^* = 53.3\ kPa$ the throat pressure is equal to the receiver pressure. Thus

$$\frac{p}{p_o} = \frac{80}{101} = 0.792$$

Using isentropic tables. The closest value of p/po = 0.79012. We will use isentropic tables without interpolation.

For p/po = 0.790 approximately to the above value

$$M = 0.59, \frac{T}{T_o} = 0.935, \frac{\rho}{\rho_o} = 0.8451$$

Therefore, T = 0.935 × 288 = 269.28 K

and $\rho = 0.8451 \times 1.23$

$$= 1.03947\ kg/m^3$$

$$a = \sqrt{k RT}$$

$$= \sqrt{1.4 \times 287 \times 269.28}$$

= 328.9326 m/s

V = M a

= 0.59 × 328.9326

= 194.07 m/s

$$\dot{m} = \rho A V$$

$= 1.03947 \times 1 \times 10^{-4} \times 194.07$

$= 0.020173$ kg/s Ans

(b) For $p_{receiver} = 40$ kPa < 55.3 kPa = p^*, the throat pressure is equal to 53.3 kPa and the duct is choked with M = 1. From isentropic tables for M = 1, we get

$$\frac{T}{T_o} = 0.8333, \quad \frac{\rho}{\rho_o} = 0.6339$$

Therefore, $T = 0.8333 \times 288 = 239.99 \approx 240\ K$

and $\rho = 0.6339 \times 1.23 = 0.779\ kg/m^3$

$a = \sqrt{k RT} = \sqrt{1.4 \times 287 \times 240} = 310.535$ m/s

$\dot{m} = \rho A V = 0.779 \times 1 \times 10^{-4} \times 310.535$

$= 0.02419\ kg/s$ Ans

3.3 Air in a tank at 700 kPa and 25°C discharges into a standard atmosphere through a converging nozzle. For isentropic flow

(a) What is the exit pressure ?

(b) What is the air temperature at the nozzle exit ?

(c) What nozzle exit diameter will discharge 0.8 kg/s ?

Solution The standard atmospheric pressure is 101.3 kPa.

Therefore, $\frac{101.3}{700} = 0.14 < 0.528$

So exit flow is sonic.

(a) $\frac{p_e}{p_o} = 0.528$

$\therefore$ $p_e = 0.528 \times 700 = 370$ kPa Ans

(b) $T_e = \frac{5}{6} T_o = \frac{5}{6} \times 298 = 248\ K$ Ans

(c) $\rho_e = \frac{370}{0.287 \times 248} = 5.18\ kg/m^3$

$a_e = \sqrt{k RT_e} = \sqrt{1.4 \times 287 \times 248} = 316\ m/s$

$A_e = A^* = \frac{\dot{m}}{\rho_e a_e} = \frac{0.8}{316 \times 5.18} = 4.89 \times 10^{-4}\ m^2$

$D_e = \sqrt{\frac{4A_e}{\pi}} = \sqrt{\frac{4 \times 4.89 \times 10^{-4}}{\pi}} = 25\ mm$ Ans

3.4 Air flows through a convergent nozzle at a stagnation pressure and temperature of 300 kN/m^2 and 400 K respectively. The exit area of the nozzle is 0.1 m^2. Find the mass flow rate, exit temperature, pressure and Mach number for discharge into an ambient of (a) 200 kN/m^2 (b) 100 kN/m^2..

Solution

$$\left(\frac{p_o}{p}\right)_{critical} = \left[1 + \frac{k-1}{2}\right]^{\frac{k}{k-1}}$$ Putting M = 1.

Check that $\left(\frac{p_o}{p_a}\right) > \left(\frac{p_o}{p}\right)_{critical}$

If not then p = p_a

If $\frac{p_o}{p_a} > \left(\frac{p_o}{p}\right)_{critical}$ then the nozzle is choked, the pressure in the exit plane is the critical value and not p_a, the exit M number is unity and this value may then be used.

$$\left(\frac{p_o}{p}\right)_{critical} = \left[1 + \frac{1.4-1}{2}\right]^{\frac{1.4}{0.4}} = 1.893$$

$$\frac{p_o}{p_a} = \frac{300}{200} = 1.5 < \left(\frac{p_o}{p}\right)_{critical}$$

therefore $p_{exit} = p_a$

$$\therefore \quad 1.5 = \left[1 + \frac{1.4-1}{2}M^2\right]^{3.5}$$

$$M = 0.784$$

$$\therefore \quad T_o = (1.5)^{1/3.5} = 1.12$$

$$\therefore \quad T = 357 \text{ K}$$

$$\dot{m} = \frac{0.1 \times 300}{400^{1/2}}\left(\frac{1.4 \times 10^3}{0.287}\right)^{1/2}$$

$$\times 0.784\left[1 + \frac{0.4}{2}(0.784)^2\right]^{-\frac{2.4}{0.8}}$$

$$\dot{m} = 57.9 \ kg/s$$

(b) $\dfrac{p_o}{p_a} = \dfrac{300}{100} = 3 > \left(\dfrac{p_o}{p}\right)_{critical}$

Hence the nozzle is choked. and $p = \dfrac{300}{1.893}$

∴ $p = 158\ \text{kN/m}^2$ and $M = 1$.

and $T = \dfrac{T_o}{\left[1 + \dfrac{0.4}{2}\right]} = \dfrac{400}{1.2} = 333\ K.$

$$\dot{m} = \frac{0.1 \times 300}{(400)^{1/2}} \left(\frac{1.4 \times 10^3}{0.287}\right)^{1/2} \times 1 \times (1.2)^{-3}$$

$= 60.6$ kg/s Ans

3.5 Air flows through an adiabatic frictionless passage. At station one, the Mach number $M_1 = 0.9$, and the static pressure is 4 bar. At station two the Mach number $M_2 = 0.2$. Calculate the change in the static pressure between station one and two.

Solution From the isentropic tables

$$At\ M_1 = 0.9, \quad \frac{p_1}{p_{o1}} = 0.59126$$

$$At\ M_2 = 0.2, \quad \frac{p_2}{p_{o2}} = 0.97250$$

Here $p_{o1} = p_{o2}$

$$\therefore \quad \frac{p_2}{p_1} = \left(\frac{p_2/p_{o2}}{p_1/p_{o1}}\right)$$

$$= \frac{0.97250}{0.59126} = 1.6448$$

$$\therefore \quad p_2 = 4 \times 1.6448 = 6.5792\ bar$$

$p_2 - p_1 = 6.5792 - 4 = 2.5792\ bar$ Ans

3.6 Air flows isentropically in a channel. At section one, the Mach number is 0.3, the area is 0.001 m^2, and the absolute pressure and the temperature are 650 kPa and 62°C, respectively. At section two, the Mach number is 0.8. Evaluate the properties at section two.

Solution To accelerate a subsonic flow requires a converging nozzle.

Stagnation conditions remain fixed for isentropic flow, therefore

$$T_{o2} = T_{o1} = T_1\left[1 + \frac{k-1}{2}M_1^2\right]$$

$$= 335\,[1 + 0.2\,(0.3)^2] = 341.03 \text{ K.}$$

$$T_2 = \frac{T_{o2}}{\left[1 + \frac{k-1}{2}M_2^2\right]}$$

$$= \frac{341.03}{[1 + 0.2\,(0.8)^2]} = \frac{341.03}{1.128} = 302.33\ K. \qquad \text{Ans}$$

$$a_2 = \sqrt{k\,RT_2} = \sqrt{1.4 \times 287 \times 302.33} = 348.535\ m/s \qquad \text{Ans}$$

$$V_2 = M_2\,a_2 = 0.8 \times 348.535 = 278.828\ m/s \qquad \text{Ans}$$

Using isentropic relation

$$\frac{p_2}{p_1} = \left(\frac{T_2}{T_1}\right)^{k/k-1} = \left(\frac{302.33}{335}\right)^{3.5} = 0.69827 \qquad \text{Ans}$$

$$p_2 = 0.69827 \times 650 = 453.88\ kPa \qquad \text{Ans}$$

Also, $$\rho_2 = \frac{p_2}{RT_2} = \frac{453.88}{0.287 \times 302.33} = 5.23\ kg/m^3 \qquad \text{Ans}$$

$$\rho_1 = \frac{p_1}{RT_1} = \frac{650}{0.287 \times 335} = 6.76\ kg/m^3$$

$$a_1 = \sqrt{k\,RT_1} = \sqrt{1.4 \times 287 \times 335} = 366.88\ m/s$$

$$V_1 = M_1\,a_1 = 0.3 \times 366.88 = 110.0648\ m/s$$

Using the continuity equation

$$\dot{m} = \rho_1 A_1 V_1 = \rho_2 A_2 V_2 = constant$$

$$A_2 = \frac{\rho_1}{\rho_2} \times A_1 \times \frac{V_1}{V_2}$$

$$= \frac{6.76}{5.23} \times 0.001 \times \frac{110.0648}{278.825}$$

$$= 5.102 \times 10^{-4}\ m^2 \qquad \text{Ans}$$

Thus $A_2 < A_1$, as expected, Finally

$$p_{o2} = p_2\left[1 + \frac{k-1}{2}M_2^2\right]^{k/k-1}$$

$$= 453.88\,[1 + 0.2\,(0.8)^2]^{3.5} = 691.867\ kPa \qquad \text{Ans}$$

Stagnation pressure is constant for isentropic flow therefore $p2_o = p_{o1}$

Aliter :— The problem can be solved very easily with the help of isentropic flow tables.

M	T/T_o	p/p_o	ρ/ρ_o	A/A^*
03.	0.9823	0.9395	0.9564	2.035
0.8	0.8865	0.6560	0.7400	1.038

For isentropic flow $T_{o1} = T_{o2} = T_o$ *Thus*

$$\frac{T_2}{T_1} = \frac{T_2}{T_o} \times \frac{T_o}{T_1} = \frac{(T/T_o)_2}{(T/T_o)_1} = \frac{0.8865}{0.9823} = 0.9025$$

$$T_2 = 0.9025 \times 335 = 302.3375\ K \qquad \text{Ans}$$

Also $p_{o2} = p_{o1} = p_o$ *so*

$$\frac{p_2}{p_1} = \frac{p_2}{p_o} \times \frac{p_o}{p_1} = \frac{(p/p_o)_2}{(p/p_o)_1} = \frac{0.6560}{0.9395} = 0.6982$$

$$p_2 = 0.6982 \times 650 = 453.83\ k\,Pa \qquad \text{Ans}$$

$$\rho_2 = p_2/RT_2 = 5.23 \text{ calculated earlier}$$

The stagnation properties are

$$T_{o2} = T_{o1} = T_1/(T/T_o)_1 = \frac{335}{0.9823} = 341.036\ K \qquad \text{Ans}$$

$$p_{o2} = p_{o1} = p_1/(p/p_o)_1 = \frac{650}{0.9395} = 691.85\ kPa \qquad \text{Ans}$$

The area may be computed using the ratio A/A^*.

Thus, since A^* = Constant,

$$\frac{A_2}{A_1} = \frac{A_2}{A^*} \times \frac{A^*}{A_1} = \frac{(A/A^*)_2}{(A/A^*)_1} = \frac{1.038}{2.035} = 0.5100737$$

$$A_2 = 0.5100 \times 0.001 = 5.1 \times 10^{-4}\ m^2 \qquad \text{Ans}$$

3.7 Air is expanded in a nozzle from 20 bar, 200°C, to 2 bar. The mass rate of flow through the nozzle is 5 kg/s. Assume the flow to be reversible and adiabatic.

(a) Determine density, velocity, Mach number, and cross sectional area.

(b) Determine the throat and exit areas for the nozzle.

(c) What back pressure will be required to cause subsonic flow throughout the entire nozzle with M = 1 at the throat ?

(d) Determine the mass flow rate through the nozzle for a back pressure of 1.9 bar.

Solution

(a) $p^* = 20 \times 0.5283 = 10.566\ bar$

$T^* = (200 + 273) \times 0.8333 = 394.165\ K$

$a^* = \sqrt{k\ RT^*} = \sqrt{1.4 \times 287 \times 394.165} = 397.964\ m/s$

$p^* = \rho^* RT^*$

$10.566 \times 100 = \rho^* \times 0.287 \times 394.165$

$e^* = 9.34\ kg/m^3$

$v^* = \frac{1}{\rho^*} = 0.1070\ m^3/kgm$

$\dot{m} = \rho^* A^* a^*$

$5 = 9.34 \times A^* \times 397.964$

$A^* = 1.345 \times 10^{-3}\ m^2$ Ans

(b) For $\frac{p_e}{p_o} = \frac{2}{20} = 0.1$

From the isentropic tables,

$M^* = 1.701, \frac{T_e}{T_o} = 0.5176, \frac{A_e}{A^*} = 1.918$

$M^* = \frac{V_e}{a^*}$

$1.701 = \frac{V_e}{397.964}$

$V_e = 676.936\ m/s$ Ans

$T_e = 0.5176 \times 473 = 244.8248\ K$

and $A_e = 1.918 \times 1.345 \times 10^{-3} = 2.5797 \times 10^{-3}\ m^2$ Ans

(c) For the subsonic flow at $\frac{A_e}{A_*} = 1.918$

Using isentropic tables

$M_e = 0.3236$

$p_e/p_o = 0.92912 \therefore p_2 = 0.92912 \times 20 = 18.58\ bar$

(d) $\dfrac{p_e}{p_{ox}} = \dfrac{1.9}{2} = 0.95$

From isentropic tables

$$M_e \approx 0.27, \frac{T_e}{T_{ox}} = 0.9855,$$

Therefore, $T_e = 0.9855 \times 473 = 466.1415\ K$

$$a_e = \sqrt{k\,RT_e}$$

$$= \sqrt{1.4 \times 287 \times 466.1415} = 432.7766\ m/s$$

$$M_e = \frac{V_e}{a_e}$$

$$\therefore \quad V_e = 0.27 \times 432.7766 = 116.85\ m/s$$

$$p_e = \rho_e RT_e$$

$$1.9 \times 100 = \rho_e \times 0.287 \times 466.1415$$

$$\rho_e = 1.42\ kg/m^3$$

$$\dot{m}_e = \rho_e A_e V_e$$

$$= 1.42 \times 2.5797 \times 10^{-3} \times 116.85 = 0.428\ kg/s$$ Ans

3.8 Air enters a duct with a Mach number of 2.5 and a static pressure of 0.35 bar. The inlet area is 0.1 m^2. The Mach number at the exit of the duct is 1 and the exit area is 0.15 m^2. What is the total pressure change between the inlet and exit of the duct. Take k = 1.4.

Solution

From table at M = 2.5

$$\frac{p_1}{p_{o1}} = 0.05853$$

$$\therefore \quad p_{o1} = \frac{p_1}{0.05853} = \frac{0.35}{0.05853} = 5.98\ bar$$

$$\frac{A_1}{A^*} = 2.6367$$

It is a case of non isentropic flow.

$$\frac{p_{o1}}{p_{o2}} = \frac{A_2^*}{A_1^*} = \frac{\left(\frac{A_1}{A_1^*}\right)_{M_1}}{\left(\frac{A_2}{A_2^*}\right)_{M_2}} \times \frac{A_2}{A_1}$$

$$p_{o2} = 1 \times \frac{5.98 \times 0.1}{2.6367 \times 0.15} = 1.51 \; bar$$

$$\therefore \quad p_{o1} - p_{o2} = 5.98 - 1.51 = 4.47 \; bar$$ Ans

3.9 Air flows isentropically through a duct. At a given point the area is 500 cm^2 and the Mach number is 0.4. At another point in the duct the area is 400 cm^2. What is the Mach number at the second point ? What would be the area at a point where M = 1. Take k = 1.4.

Solution

(a) At $M_1 = 0.4, \frac{A_1}{A^*} = 1.5901$

$$\frac{A_2}{A^*} = \frac{A_2}{A_1} \times \frac{A_1}{A^*}$$

$$= \frac{400}{500} \times 1.5901 = 1.27208$$

From table $M_2 \approx 0.54$ *Subsonic*

$M_2 \approx 1.63$ *Supersonic* Ans

Both the values are possible answers because further information is not given

(b) $$\left(A^*\right)_{m=1} = \left(\frac{A^*}{A_1}\right) \times A_1$$

$$= \frac{1}{1.59} \times 500 = 314.5 \; cm^2$$ Ans

If $M_2 = 1.63$ it would be necessary that a throat exist with an area equal to A^*. If $M_2 = 0.54$ no throat need exist. The area may decrease steadily from $A_1 = 500 \; cm^2$ to $A_2 = 400 \; cm^2$.

3.10 Air flows adiabatically through pipe with a constant area. At point one the total pressure is 3 bar and the Mach number is 0.4. Further downstream the total pressure is 2 bar. What is the Mach number at the second point for a subsonic flow.

Solution

$$At\ M_1 = 0.4\ p_1 = 3\ \text{bar}$$

From table $\frac{p_1}{p_o} = 0.89562$

$$\therefore \quad p_o = \frac{3}{0.89562} = 3.3512\ bar$$

At about (2) $p_2 = 2$ bar

$$p_{o1} = p_{o2}$$

$$\frac{p_2}{p_o} = \frac{2}{3.3512} = 0.5968$$

From isentropic table, for subsonic flow, correspond to above $\frac{p_2}{p_o} = 0.5968$

M = 0.89 Ans

3.11 Air in a reservoir at 825 kPa and 50°C discharges through a converging nozzle to the atmosphere. Design a nozzle for flow rate of 1.5 kg/s. specify exit diameter.

Design a converging-diverging nozzle for full expansion at the same flow rate. Specify throat and exit diameters.

Solution

$$p_r < p^* = 0.528 \times 825 = 436\ kPa$$

so exit flow is sonic

$$T^* = \frac{5}{6} T_o = \frac{5}{6} \times 323 = 269\ K$$

$$a^* = \sqrt{k\,RT^*}$$

$$= \sqrt{1.4 \times 287 \times 269} = 328.76\ m/s$$

$$\rho^* = \frac{p^*}{RT^*} = \frac{436}{0.287 \times 269} = 5.647\ kg/m^3$$

$$A_e = \frac{\dot{m}}{\rho^*\ a^*} = \frac{1.5}{5.647 \times 328.76} = 8.079 \times 10^{-4}\ m^2$$

$$A_e = \frac{\pi}{4} D_e^2 \quad \therefore D_e = \sqrt{\frac{4A_e}{\pi}}$$

$$= \sqrt{\frac{4 \times 8.079 \times 10^{-4}}{\pi}} = 32 \ mm$$ Ans

(b) The throat diameter is same as in part (a) i.e. 32 mm, since the throat flow is sonic in each case.

$$\frac{p_e}{p_o} = \frac{101.3}{825} = 0.12279$$

From isentropic table

$$M_e = 2.026$$

$$\frac{A_e}{A^*} = 1.725$$

$$\text{so} \quad \frac{D_e}{D^*} = \sqrt{1.725}$$

$$\therefore \quad D_e = 32\sqrt{1.725} = 42 \ mm$$ Ans

3.12 Air is drawn isentropically from a standard atmosphere at sea level (101.3 kPa and 15°C) through a converging diverging nozzle. Static pressure at two different locations are 80 kPa and 40 kPa, respectively.

(a) What is the Mach number at each of these locations ?

(b) What is the velocity at each of these locations ?

Solution

(a) $$\frac{p_1}{p_o} = \frac{80}{101.3} = 0.79; \ M_1 = 0.59$$ Ans

$$\frac{p_2}{p_o} = \frac{40}{101.3} = 0.395; \ M_2 = 1.232$$ Ans

(b) $$\frac{T_1}{T_o} = 0.9349 \ \textit{from isentropic table}$$

$$T_1 = 0.9349 \times 288 = 269.25 \ K$$

$$a_1 = \sqrt{kRT_1} = \sqrt{1.4 \times 287 \times 269.25} = 328.915 \ m/s$$

$$M_1 = \frac{V_1}{a_1}$$

$$0.59 = V_1/328.915$$

$$V_1 = 0.59 \times 328.915 = 194 \ m/s$$ Ans

$$\frac{T_2}{T_o} = 0.767 \ \textit{from isentropic table}$$

$$T_2 = 0.767 \times 288 = 220.896\ K$$

$$a_2 = \sqrt{k\,RT_2}$$

$$= \sqrt{1.4 \times 287 \times 220.896} = 297.92\ m/s$$

$$M_2 = \frac{V_2}{a_2}$$

$$1.232 = \frac{V_2}{297.92}$$

$$V_2 = 1.232 \times 297.92 = 367\ m/s$$

Ans

3.13 Air flows steadily and isentropically through a converging diverging nozzle. At the throat the air is at 140 kPa absolute, and 60°C. The throat cross sectional area is 0.05 m^2. At a certain section in the diverging part of the nozzle, the pressure is 70 kPa absolute. Calculate the speed and area at this section.

Solution

Since $p_1 < p_t$, flow downstream of throat is supersonic, therefore $M_t = 1$.

$$T_o = T_t\left(1 + \frac{k-1}{2}M^2\right)$$

$$= 333\,[1 + 0.2\,(1)^2]$$

$$= 399.6\ K$$

$$p_o = p\left(1 + \frac{k-1}{2}M^2\right)^{k/k-1} = 140\,(1.2)^{3.5} = 265\ kPa$$

$$a_t = \sqrt{k\,RT_t} = \sqrt{1.4 \times 287 \times 333} = 365.786\ m/s$$

At section x

$$\frac{p_o}{p_x} = \left[1 + \frac{k-1}{2}M_x^2\right]^{k/k-1}$$

$$\left(\frac{265}{70}\right)^{1/3.5} = 1 + 0.2\,M_x^2$$

$$1.46 = 1 + 0.2\,M_x^2$$

$$M_x^2 = 2.314\ \therefore\ M_x = 1.52$$

$$T_x = \frac{T_o}{1 + \frac{k-1}{2}M_x^2} = \frac{399.6}{1 + 0.2\,(1.52)^2} = \frac{399.6}{1.4628} = 273.17\ K.$$

$$V_x = M_x\,a_x = 1.52 \times \sqrt{1.4 \times 287 \times 273.17}$$

$= 503.58$ m/s Ans

$$\dot{m} = \rho_x V_x A_x = \rho_t V_t A_t$$

Therefore, $A_x = \dfrac{\rho_t}{\rho_x} \times \dfrac{V_t}{V_x} \times A_t$

$$= \left(\frac{p_t}{p_x}\right)^{1/k} \times \frac{V_t}{V_x} \times A_t$$

$$= \left(\frac{140}{70}\right)^{1/1.4} \times \frac{365.786}{503.58} \times 0.05$$

$= 1.64 \times 0.726 \times 0.05 = 0.059586\ m^2$ Ans

Aliter :— Using isentropic tables

$$At\ M_t = 1,\ \frac{T_t}{T_o} = 0.8333,\ \frac{p_t}{p_o} = 0.5283$$

$\therefore \quad T_o \approx 400\ K,\ p_o \approx 265\ kPa$

For $\quad p_x/p_o = 0.2642\ ,\ M_x = 1.52$

$T_x/T_o = 0.684$ and $A_x/A^* = 1.19$

$\therefore \quad T_x \approx 274\ K,\ A_x = 0.0595\ m^2$ Ans

3.14 Air flows isentropically in a converging diverging nozzle, with exit area of 0.001 m^2. The nozzle is fed from a large plenum where the stagnation temperature and the stagnation pressure are 350 K and 10 bar respectively. The exit pressure is 954 kPa, and the Mach number at the throat is 0.68. Flow conditions at the throat and the exit Mach number are to be determined.

The above nozzle has a design back pressure of 72.8 kPa but is operated at a back pressure of 50 kPa. Flow within the nozzle may be assumed isentropic. Determine the exit Mach number and mass flow rate.

Solution

The stagnation properties are constant for isentropic flow. At M = 0.68 from isentropic tables

$$\frac{T}{T_o} = 0.9154,\ \frac{p}{p_o} = 0.7738,\ \frac{\rho}{\rho_o} = 0.8016$$

and $\quad A/A^* = 1.11$

$$T_t = 0.9154\ T_o = 0.9154 \times 350 = 320.39\ K$$

$$P_t = 0.7338\, p_o = 0.7338 \times 1000 = 733.8\ kPa$$

$$\rho_t = 0.8016\, \rho_o$$

$$\rho_t = 0.8016 \frac{p_o}{RT_o} = 0.8016 \times \frac{10000}{0.287 \times 350}$$

$$= 7.98\ kg/m^3$$ Ans

$$A = 1.11\, A^*$$

But at this point A^* is not known.

At this exit, $P_e = 954$ kPa

Thus $p_e/p_o = 0.954$

From isentropic flow tables

$$M_e = 0.26$$ Ans

Since A_e is unknown, we can compute A^*.

From isentropic flow tables tat M = 0.26,

$$A/A^* = 2.317 \text{ Thus}$$

$$A^* = \frac{A_e}{2.317} = \frac{0.001}{2.317} = 4.3159 \times 10^{-4}\ m^2$$

and $$A_t = 1.11\, A^* = 1.11 \times 4.3159 \times 10^{-4}$$

$$= 4.79 \times 10^{-4}\ m^2$$ Ans

From isentropic flow tables, $p/p_o = 0.0728$,

$$M_e = 2.36$$ Ans

Also,

$$\frac{T_e}{T_o} = 0.4731$$

$$\therefore \quad T_e = 0.4731 \times 350 = 165.585\ K$$

$$a_e = \sqrt{k\, RT_e} = \sqrt{1.4 \times 287 \times 165.585} = 257.938\ m/s$$

$$V_e = M_e\, a_e = 2.36 \times 257.938 = 608.73\ m/s$$

$$\rho_e = p_e/RT_e = \frac{72.8}{0.287 \times 165.585} = 1.53\ kg/m^3$$

$$\dot{m} = \rho_e A_e V_e = 1.53 \times 0.001 \times 608.73$$

$$= 0.9325\ kg/s$$ Ans

3.15 At a point upstream of the throat in a converging-diverging nozzle, the air speed is 172.14 m/s; pressure is 200 kPa and temperature is 22°C. The flow is isentropic and is supersonic at the nozzle exit. If the nozzle throat area is 0.01 m², determine the flow rate.

Solution

$$a_1 = \sqrt{k\,RT_1} = \sqrt{1.4 \times 287 \times 295} = 344.28$$

$$M_1 = \frac{V_1}{a_1} = \frac{172.14}{344.28} = 0.5$$

$$\frac{T_o}{T} = 1 + \frac{k-1}{2} M_1^2 = 1 + 0.2\,(0.5)^2 = 1.05$$

$$T_o = 295 \times 1.05 = 309.75\ K$$

$$\frac{p_o}{p} = \left(\frac{T_o}{T}\right)^{k/k-1} = (1.05)^{3.5} = 1.1862$$

$$p_o = 200 \times 1.1862 = 237.24\ kPa$$

Since $M_2 > 1$, the flow is choked and $M_t = 1$.

Therefore, $$\frac{T_o}{T_t} = 1 + \frac{k-1}{2} M^2 = 1 + 0.2\,(1)^2 = 1.2$$

$$T_t = \frac{T_o}{1.2} = \frac{309.75}{1.2} = 258.125\ K$$

and $$\frac{p_o}{p_t} = \left(\frac{T_o}{T_t}\right)^{k/k-1} = (1.2)^{3.5} = 1.8929$$

$$p_t = \frac{p_o}{1.8929} = \frac{237.24}{1.8929} = 125.3295\ kPa$$

$$p_t = \rho_t RT_t$$

$$125.3295 = \rho_t \times 0.287 \times 258.125$$

$$\rho_t = 1.69\ kg/m^3$$

$$a_t = \sqrt{k\,RT} = \sqrt{1.4 \times 287 \times 258.125} = 322\ m/s$$

$$V_t = M_t\,a_t = 1 \times 322 = 322\ m/s$$

$$\dot{m} = \rho_t A_t V_t$$

$$= 1.69 \times 0.01 \times 322$$

$$= 5.4426\ kg/s$$ Ans

Aliter At M = 0.5 from isentropic tables

$$\frac{T_1}{T_o} = 0.9524 \text{ and } \frac{p_1}{p_o} = 0.8430$$

$\therefore \quad T_o = 310\,K,\ p_o = 237\,kPa$

For $\quad M_t = 1\ \dfrac{p_t}{p_o} = 0.5283,\ \dfrac{T_t}{T_o} = 0.8333$

$\therefore \quad T_t = 258\,K$ and $p_t = 125\,kPa$

For the remaining procedure see above.

3.16 A small rocket motor fueled with hydrogen and oxygen is tested on a thrust stand at a simulated altitude of 10 km. The motor is operated at chamber stagnation condition of 1500 K and 8 MPa gauge. The combustion product is water vapour, which may be tested as an vapour, which may be tested as an ideal gas. Expansion occurs through a converging diverging nozzle with design Mach number of 3.5 and exit area of 700 mm^2. Evaluate the pressure at the nozzle exit plane. Calculate the mass flow rate of exhaust gas. Determine the force exerted by the rocket motor on the thrust stand. Take k = 1.3 and R = 461 J/kg K.

Solution

The pressure is 26.5 KPa at 10 km altitude.

$$P_o = 8 \text{ MPa} + 0.1 \text{ MPa} = 8.1 \text{ M Pa abs}$$

Design pressure at exit

$$\frac{p_o}{p_d} = \left[1 + \frac{k-1}{2} M^2\right]^{k/k-1} = [\,1 + 0.15\,(3.5)^2\,]^{4.33}$$

$$= (2.8375)^{4.33} = 91.7426$$

$$p_d = \frac{8.1 \times 1000}{91.7426} = 88.29\ kPa\ abs.$$

Since $p_b < p_d$, therefore, $p_b = p_d = 88.29\ kPa$

$$T_e = \frac{T}{1 + \frac{k-1}{2} M^2} = \frac{1500}{1 + 0.15\,(3.5)^2} = 528.634K$$

$$\rho_e = \frac{p_e}{RT_t} = \frac{88.29}{0.461 \times 528.634} = 0.362\ kg/m^3$$

$$a_e = \sqrt{k\,RT_e} = \sqrt{1.3 \times 461 \times 528.634} = 562.86$$

$$V_e = M\,a_e = 3.5 \times 562.86 = 1970\ m/s$$

$$\dot{m} = \rho_e V_e A_e = 0.362 \times 1970 \times \frac{700}{10^6} = 0.499\ kg/s. \qquad \text{Ans}$$

The force on test stand.

$$F_x = -R_x = -\dot{m} \quad V_e - A_e (p_e - p_b)$$

$$= -0.499 \times 1970 - \frac{700}{10^6} (88.29 - 26.5) \times 1000$$

$$= -983.423 - 43.253 = -1026.676\ N\ (to\ left)$$ Ans

3.17 A liquid motor is fueled with hydrogen and oxygen. The chamber temperature and absolute pressure are 3300 K and 6.9 MPa, respectively. The nozzle is to be designed to expand the exhaust gases isentropically to a design back pressure corresponding to an altitude to 10 km on a standard day. The thrust produced by the motor is to be 100 kN at the design conditions. Treat the exhaust gases as water vapour and assume ideal gas behaviour. Determine the propellant mass flow rate needed to produce the desire thrust, the nozzle exit area, and the area ratio, A_e/A_t. Take k = 1.3 and R = 462 J/kg K.

<u>Solution</u>

$$\frac{p_o}{p} = \left[1 + \frac{k-1}{2} M_e^2\right]^{k/k-1}$$

The back pressure is $p_e = p_b = 26.4$ kPa. at 10 km altitude.

$$\text{Therefore, } \left(\frac{690}{2.64}\right)^{0.3/1.3} = 1 + \frac{0.3}{2} M_e^2$$

$$3.5972 = 1 + 0.15\, M_e^2$$

$$M_e^2 = 17.3146$$

$$M_e = 4.16$$

$$\frac{T_o}{T_e} = 1 + \frac{k-1}{2} M_e^2$$

$$T_e = \frac{T_o}{1 + \frac{k-1}{2} M_e^2} = \frac{3300}{1 + 0.15\,(4.16)^2} = \frac{3300}{3.59584}$$

$$= 917.72 \text{ K}$$

$$V_e = M_e a_e = 4.16 \times \sqrt{k\,RT_e}$$

$$= 4.16 \times \sqrt{1.3 \times 462 \times 917.72}$$

$$= 416 \times 742.416$$

$$= 3088.45 \text{ m/s}$$

$$R_x = \dot{m} \quad V_e$$

$$100000 = \dot{m} \quad \times 3088.45$$

$$\dot{m} \quad = 32.378\ kg/s$$ Ans

$$\rho_e = \frac{p_e}{RT_e} = \frac{26.4 \times 1000}{462 \times 917.72} = 0.06226\ kg/m^3$$

$$\dot{m} \quad = \rho_e V_e A_e$$

$$32.378 = 0.06226 \times 3088.46 \times A_e$$

$$A_e = 0.168\ m^2$$ Ans

Since $M_t = 1$, then

$$T_t = \frac{T_o}{1 + \frac{k-1}{2} M^2} = \frac{3300}{1 + 0.15} = \frac{3300}{1.15} = 2869.56\ K$$

$$p_t = \frac{p_o}{\left[1 + \frac{k-1}{2} M^2\right]^{k/k-1}}$$

$$= \frac{6.9}{(1 + 0.15)^{4.33}} = \frac{6.9}{(1.15)^{4.33}} = \frac{6.9}{1.832} = 3.765\ \text{MPa}$$

$$\rho_t = p_t/RT_t = \frac{3.765 \times 1000}{0.462 \times 2869.56} = 2.84\ kg/m^3$$

$$V_t = M_t a_t = 1 \times \sqrt{k RT_t}$$

$$= \sqrt{1.3 \times 462 \times 2869.56} = 1312.8\ m/s$$

$$\dot{m} \quad = \rho_e V_e A_e = \rho_t V_t A_t \therefore \frac{A_e}{A_t}$$

$$= \frac{\rho_t}{\rho_e} \times \frac{V_t}{V_e} = \frac{2.84}{0.06226} \times \frac{1312.8}{308.845}$$

$$\therefore A_e/A_t = 19.3896$$ Ans

3.18 A Converging diverging nozzle has an exit area to throat area ratio of 2. Air enters this nozzle with a stagnation pressure of 1000 kPa and a stagnation temperature of 360 K. The throat area is 500 mm^2. Determine the mass rate of flow, exit pressure, exit temperature, exit Mach number, and exit velocity for the following conditions.

(a) Sonic velocity at the throat, diverging section acting as a nozzle.

(b) Sonic velocity at the throat, diverging section acting as a diffuser.

Solution

In the table there are two Mach number listed for $\frac{A}{A^*} = 2$ one is greater than unity and one is less than unity. When the diverging section acts as a supersonic nozzle we use the value of M > 1.

$$\therefore \quad \frac{A_e}{A^*} = 2, M_e = 2.197, \frac{p_e}{p_o} = 0.0939, \frac{T_e}{T_o} = 0.5089$$

$$\therefore \quad p_e = 0.0939\,(1000) = 93.9\ kPa$$

$$T_e = 0.5089\,(360) = 183.2\ kPa$$

$$a_e = \sqrt{k\,RT_e} = \sqrt{1.4 \times 287 \times 183.2} = 271.3\ m/s$$

$$V_e = M_e\,a_e = 2.197 \times 271.3 = 596.1\ m/s$$

The mass rate of flow can be calculated either by exit or at throat.

For (b) Part

$$\frac{A_e}{A^*} = 2, M_e = 0.308, \frac{p_e}{p_o} = 0.936,$$

$$\frac{T_e}{T_o} = 0.9812$$

$$p_e = 0.936\,(1000) = 936\ kPa$$

$$T_e = 0.9812\,(360) = 353.3\ K$$

$$a_e = \sqrt{k\,RT_e} = \sqrt{1.4 \times 287 \times 353.4} = 376.8\ m/s$$

$$V_e = M_e\,a_e = 0.308 \times 376.8 = 116\ m/s$$

Since M = 1 at the throat the mass rate of flow is the same as in (a), which is also equal to the flow in the convergent nozzle as in the ex. 3. when it is choked.

3.19 A concial order diffuser has an intake area of 0.11 m^2 and an exit area of 0.44 m^2. Air enters the diffuser with a static pressure of 0.18 MPa, static temperature of 37°C, and velocity of 267 m/s. Calculate (a) the mass flow rate of air through the diffuser, (b) the Mach number, static temperature, and static pressure of air leaving the diffuser, and (c) the net thrust acting upon the diffuser, assuming that its outer surfaces are wetted by atmospheric pressure at 0.1 MPa.

Solution

$$\dot{m} = \rho A V = \frac{p_1}{RT_1} A_1 V_1$$

$$= \frac{0.18 \times 10^3}{0.287 \times 310} \times 0.11 \times 267 = 59.42\ kg/s$$

$$a_1 = \sqrt{k\,RT_1} = 20.045\sqrt{310} = 352\ m/s$$

$$M_1 = \frac{V_1}{a_1} = \frac{267}{352} = 0.76$$

From the table $\frac{A_1}{A^*} = 1.0570,\ \frac{p_1}{p_{o1}} = 0.68207,\ \frac{T_1}{T_{o1}} = 0.89644$

$$\frac{F_1}{F^*} = 1.0284\ \therefore\ \frac{A_2}{A_1} = \frac{A_2}{A^*}\ \frac{A_1}{A^*}$$

$$\frac{A_2}{A^*} = \frac{A_2}{A_1} \times \frac{A_1}{A^*} = 4 \times 1.0570 = 4.228$$

From isentropic table

At $\frac{A_2}{A^*} = 4.228$

$$M_2 \approx 0.135,\ \frac{p_2}{p_{o2}} = 0.987,\ \frac{T_2}{T_{o2}} = 0.996$$

$$\frac{F_2}{F^*} = 3.46$$

$$\therefore\quad \frac{p_2}{p_1} = \frac{p_2/p_{o2}}{p_1/p_{o1}} = \frac{0.987}{0.682} = 1.447$$

$$\therefore\quad p_2 = 1.447 \times 0.18 = 0.26\ MPa$$

$$\frac{T_2}{T_1} = \frac{T_2/T_{o2}}{T_1/T_{o1}} = \frac{0.996}{0.89644} = 1.111$$

$$\therefore\quad T_2 = 1.111 \times 310 = 344.4\ K$$

$$F_1 = p_1 A_1 + \rho_1 A_1 V_1^2 = p_1 A_1 \left[1 + \frac{1}{RT_1} V_1^2\right]$$

$$= p_1 A_1 (1 + K M_1^2)$$

$$= 0.18 \times 10^3 \times 0.11\,(1 + 1.4 \times 0.76^2) = 35.82\ kN$$

Internal thrust $= F_2 - F_1 = 3.364\,F_1 - F_1$

$$= 2.364 \times F_1 = 2.364 \times 35.82 = 84.68\ kN$$

External thrust $= = p_o\,(A_2 - A_1)$

$= 0.1 \times 10^3 (0.44 - 0.11) = 33\ kN$

Net thrust = 84.68 – 33 = 51.68 kN

3.20 A conical diffuser has entry and exit diameters are 150 mm and 300 mm respectively. The pressure, temperature, and air velocity at entry are 69 kPa, 67°C and 180 m/s respectively. Determine exit pressure, exit velocity and the force exerted on the diffuser walls. Assume isentropic flow, Take k = 1.4 and cp = 1 kJ/kg K.

Solution

$$A_1 = \frac{\pi}{4} d_1^2 = \frac{\pi}{4}(0.15)^2 = 0.01767\ m^2$$

$$A_2 = \frac{\pi}{4} d_2^2 = \frac{\pi}{4}(0.3)^2 = 0.07068\ m^2$$

$$T_{o1} = T_1 + \frac{V_1^2}{2\,c_p} = 340 + \frac{180^2}{2 \times 1 \times 1000} = 356.2\ K$$

$$\frac{T_1}{T_{o1}} = \frac{340}{356.2} = 0.952$$

From isentropic flow tables

$$At\ \frac{T_1}{T_{o1}} = 0.952,$$

$$\frac{p_1}{p_{o1}} = 0.843\ , M_1 = 0.5\ , \frac{A_1}{A_1^*} = 1.34,$$

$$F_1/F_1^* = 1.203$$

$$\text{Therefore, } \frac{p_1}{p_{o_1}} = \frac{0.69}{p_{o_1}} = 0.843$$

$$p_{o1} = p_{o2} = \frac{0.69}{0.843} = 0.818\ bar$$

$$A_1^* = A_2^* = 0.01767/1.34 = 0.01318\ m^2$$

$$p_1^* = 0.528\ p_{o1} = 0.528 \times 0.818 = 0.432\ bar$$

$$F_1^* = F_2^* = p_1^*\ A_1^*\ (1 + k)$$

$= 0.432 \times 100 \times 0.01318\ (1 + 1.4)$

$= 1.367$ kN

$$\frac{A_2}{A_2^*} = \frac{0.07068}{0.01318} = 5.36267 \approx 5.37$$

From isentropic tables.

$$At\ A_2/A_2^* = 5.37$$

$$M_2 = 0.107\,, \frac{p_2}{p_o^2} = 0.992\,, \frac{F_2}{F_2^*} = 4.3$$

$$p_2 = 0.992\ p_{o_2} = 0.992 \times 0.818 = 0.81\ bar$$

Using the continuity equation

$$A_1 V_1 = A_2 V_2$$

$$V_2 = \frac{A_1}{A_2} V_1 = \frac{0.01767}{0.07068} \times 180 = 45\ m/s \qquad \text{Ans}$$

$$F = F_2 - F_1$$

$$= 4.3\ F_2^* - 1.203\ F_1^*$$

$$= 3.097\ F_1^*$$

$$= 3.097 \times 1.367 = 4.2336\ kN \qquad \text{Ans}$$

3.21 A supersonic diffuser decelerates air isentropically from a Mach number of 3 to a Mach number of 1.4. If the static pressure at the diffuser inlet is 30 kPa, Calculate the static pressure rise in the diffuser and the ratio of inlet to outlet area of the diffuser.

Solution

$$\frac{p_2}{p_1} = \frac{p_2}{p_{o2}} \times \frac{p_{o1}}{p_1} = \frac{p_2}{p_o} \times p_{o1}$$

Here $p_{o1} = p_{o2} = p_o = constant$

and $p_o/p = \left[1 + \frac{k-1}{2} M^2\right]$

Therefore,

$$\frac{p_2}{p_1} = \left[\frac{1 + \frac{k-1}{2} M_1^2}{1 + \frac{k-1}{2} M_2^2}\right]^{k/k-1} = \left[\frac{1 + 0.2\,(3)^2}{1 + 0.2\,(1.4)^2}\right]^{3.5} = 11.5$$

$$p_2 = 11.5\, p_1$$

Therefore,

$$p_2 - p_1 = 11.5\,p_1 - p_1$$

$$= 10.5\,p_1 = 10.5 \times 30 = 315\ kPa \qquad \text{Ans}$$

$$\frac{A_1}{A_2} = \frac{\rho_2 V_2}{\rho_1 V_1} = \left(\frac{p_2}{p_1}\right)^{1/k}\left(\frac{V_2}{V_1}\right) = \left(\frac{p_2}{p_1}\right)^{1/k}\left(\frac{M_2}{M_1}\right)\left(\frac{a_2}{a_1}\right)$$

$$\frac{A_1}{A_2} = \left(\frac{p_2}{p_1}\right)^{1/k}\left(\frac{M_2}{M_1}\right)\sqrt{\left(\frac{T_2}{T_1}\right)}$$

$$= (11.5)^{1/1.4} \times \frac{1.4}{3} \times \sqrt{\frac{1 + 0.2\,(3)^2}{1 + 0.2\,(1.4)^2}} = 3.79 \qquad \text{Ans}$$

<u>Aliter</u> :– From isentropic tables

At $\quad M_1 = 3, \dfrac{p_1}{p_o} = 0.0272\,, \dfrac{A_1}{A^*} = 4.235$

and at $M_2 = 1.4\,, \dfrac{p_2}{p_o} = 0.3142\,, \dfrac{A_2}{A^*} = 1.115$

Therefore,

$$\frac{p_2}{p_1} = \frac{p_2}{p_o} \times \frac{p_o}{p_1} = \frac{0.3142}{0.0272} = 11.55$$

and $\quad \dfrac{A_1}{A_2} = \dfrac{A_1}{A^*} \times \dfrac{A^*}{A^2} = \dfrac{4.235}{1.115} = 3.798 \qquad$ Ans

EXERCISES

A. THEORY

3.1 Discuss with suitable sketches and graphs the effect of pressure ratio on convergent-divergent nozzle operation.

Explain what is meant by 'overexpanding' and 'under expanding' nozzle.

3.2 Describe the behaviour of flow in a convergent-divergent nozzle when it is operated at

(i) design pressure ratio

(ii) pressure ratio higher than design value and

(iii) pressure ratio lower than the design value.

3.3 Fully explain why there is a maximum rate of mass flow for a gas which is deing discharged through a convergent nozzle under steady supply conditions.

3.4 Discuss the flow characteristic of, a convergent divergent nozzle of fixed geometry.

3.5 Derive the Hugoniot equation for one dimensional isentropic flow.

3.6 Discuss the variation of pressure, density and velocity for isentropic flow.

3.7 Derive the Fliegner's formula.

3.8 Draw the flow characteristic of convergent nozzle and discuss the two regimes.

3.9 Draw the flow characteristic of convergent-divergent nozzle and discuss the four regimes.

3.10 Draw the curve between A/A^* and Mach number, M and discuss for acceleration and deceleration.

3.11 For a variable area flow, derive the relation

$$\frac{dA}{A} = \frac{dV}{V}(M^2 - 1)$$

and hence discuss the criteria for acceleration and deceleration.

3.12 Show that for variable area flow.

(a) $$a^2 = \frac{dp}{\rho V^2}(1 - M^2)$$

(b) $$\frac{d\rho}{\rho} = \frac{dA}{A}\left(\frac{M^2}{1 - M^2}\right)$$

(c) $$\frac{dA}{A} = -\frac{dV}{V}(M^2 - 1)$$

3.13 Derive the relation between area and Mach number for isentropic flow.

3.14 Derive the relation for variable area flow :

(a) $$\frac{A}{A^*} = \frac{1}{M}\left[\frac{1 + (k-1)/2\ M^2}{(k+1)/2}\right]^{(k+1)/2\,(k-1)}$$

(b) $$\frac{A}{A^*} = \frac{1}{M}\left[\left(\frac{2}{k+1}\right)\left(1 + \frac{k-1}{2}M^2\right)\right]^{(k+1)/2\,(k-1)}$$

(c) $$\frac{A}{A^*} = \frac{1}{M}\left[\left(\frac{2}{K+1}\right)\left(\frac{k-1}{k+1}\right)M^2\right]^{(k+1)/2\,(k-1)}$$

(d) $$\left(\frac{A}{A^*}\right)\left(\frac{p}{p_o}\right) = \frac{1}{M}\sqrt{\frac{(2/k+1)^{(k+1)/(k-1)}}{\left(1 + \frac{k-1}{2}M^2\right)}}$$

3.15 Prove that

$$\frac{\dot{m}}{A} = \frac{p_o}{\sqrt{T_o}} \sqrt{k/R} \frac{M}{\left(1+\frac{k-1}{2}M^2\right)^{(k+1)/2(k-1)}}$$

3.16 Prove that

$$\dot{m}_{max} = A^* \frac{p_o}{\sqrt{T_o}} \sqrt{\frac{k}{R}} \frac{1}{\left(1+\frac{k-1}{2}M^2\right)^{(k+1)/2(k-1)}}$$

B. UNSOLVED PROBLEMS

3.1 Air is flowing isentropically through a converging nozzle discharges to the atmosphere. At the section where the absolute pressure is 179 kPa, the temperature is 39°C, and the air speed is 177 m/s. Determine the nozzle throat pressure. [p_t = 112 kPa]

3.2 Air, with a stagnation pressure of 650 kPa (abs) and a stagnation temperature of 350 K, is flowing isentropically through a converging nozzle. At the section in the nozzle where the area is 2.6×10^{-3} m^2, the Mach number is 0.5. The nozzle discharges to a back pressure of 270 kPa (abs). Determine the exit area of the nozzle. [$A_t = A^* = 1.92 \times 10^{-3}$ m^2]

3.3 A large tank supplies air to a converging nozzle that discharges to atmospheric pressure. Assume the flow to be reversible and adiabatic.

(a) For what range of tank pressures will the flow at the nozzle exit be sonic i.e. M = 1 ?

(b) If the tank pressure is 600 kPa (abs) and the temperature is 600 K, what is the mass flow rate through the nozzle if the exit area is 1.29×10^{-3} m^2 ? [191 kPa, 1.28 kg/s]

3.4 Air enters a converging diverging nozzle with negligible speed at an absolute pressure of 10 bar and a temperature of 333 K. If the flow is isentropic and the exit temperature is – 11°C, what is the Mach number at exit ? [M_1 = 1.16].

3.5 A convergent nozzle with a throat area of 0.001 m^2 is operated with air at a back pressure of 591 kPa. The nozzle is fed from a large plenum chamber where the absolute stagnation pressure and the temperature are 1 MPa and 60°C, respectively. The exit Mach number and mass flow rate are to be determined using (a) isentropic flow relations and (b) isentropic flow tables. [M = 0.9, flow is not choked $\dot{m}$ = 2.2 kg/s]

3.6 A nozzle is designed to expand air isentropically to atmospheric pressure from a large tank in which properties are held constant at 5°C and 304 kPa. The desired flow rate is 1 kg/s. Determine the exit area of the nozzle. [ρ_e =

1.73 kg/m^3, V_e = 388 m/s, $A_e = 1.49 \times 10^{-3}$ m^2]

3.7 Air flows steadily and isentropically into an aircraft inlet at a rate of 100 kg/s. At the section where the area is 0.464 m^2, M = 3, t = – 60°C, and the absolute pressure is 15 kPa. Determine the speed and cross sectional area downstream where t = 138°C [610 m/s, 0.129 m^2]

3.8 Air at an absolute pressure of 60 kPa and 27°C enters a passage at 486 m/s. The cross sectional area at the entrance is 0.02 m^2. At the downstream, the pressure is 78.8 kPa absolute. Assuming isentropic flow, calculate the Mach number at downstream condition. [M_2 = 1.2]

3.9 Air is flowing isentropically in a converging diverging nozzle. At the section in the converging portion where the area is 1250 mm^2, the pressure is 600 kPa (abs), the temperature is 22°C, and the Mach number is 0.5. Determine the area in the diverging section where the Mach number is 2. [A_2 = 1563 mm^2]

3.10 Air flows through a constant area duct. At section one conditions are 550 K, 973 kPa and Mach number is 0.2. At section two flow is at 1100 K, 910 kPa and Mach number is 0.3.

(a) Is the flow isentropic ? Justify your answer.

(b) Calculate the stagnation enthoply, stagnation temperature, and stagnation pressure at sections one and two. [No, $T_{0_1} = 554\,K$, $T_{0_2} = 1120\,K$, $P_{0_1} = 1000\,kPa$, $= p_{o2} = 969\,k\,Pa$, $h_{0_1} = 554\,kJ/kg$, $h_{0_2} = 1.12\,MJ/kg$]

3.11 Consider a supersonic nozzle with a ratio of exit to throat area of 2. The nozzle is supplied with air at low speed at 670 kPa at 333 K. The overall efficiency from inlet to exit is 90 % but the flow is isentropic to the throat. Calculate pressure, velocity and Mach number at exit and compare with corresponding values for isentropic flow. [62.8 kPa, 169 K, M_e = 2]

3.12 Air flows steadily and isentropically through a passage. At the upstream section where the cross sectional area is 0.02 m^2, the air is 40 kPa (absolute), 60°C ad the Mach number is 2. At the downstream section, the speed is 519 m/s. Calculate the Mach number at the downstream section. [M_2 = 1.2].

3.13 Air flows isentropically through a converging nozzle into a receiver in which the absolute pressure is 240 kPa. The air enters the nozzle with negligence speed at a pressure of 406 kPa (abs) and a temperature of 95°C. Determine the flow rate through the nozzle for a nozzle throat area of 0.01 m^2. [$\dot{m}$ = 8.47 kg/s].

3.14 Air flows isentropically through a converging nozzle attached to a large tank where the absolute pressure is 171 kPa and the temperature is 27°C. At the inlet section the Mach number is 0.2. The nozzle discharges to the atmosphere the discharge area is 0.015 m^2. Determine the magnitude and direction of the force that must be applied to hold the nozzle in place. [A_1 = 0.0441 m^2, p_1 = 166 kPa, $\dot{m}$ = 5.92 kg/s, V_1 = 69.2 m/s, V_2 = 290 m/s, R_x =

1560 N (to the left)]

3.15 Air flows from a large tank (p = 650 kPa (abs), t = 550°C through a converging nozzle with a throat area of 600 mm^2, and discharges to the atmosphere. Determine the mass flow rate for isentropic flow through the nozzle. [$\dot{m}$ = 0.548 kg/s]

3.16 A nozzle in a wind tunnel gives a test section Mach number of 2. Air enters the nozzle from a large reservoir at 69 kPa and 37°C. The cross sectional area of the throat is 1000 cm^2. Determine the following for the tunnel.

(a) pressure, temperature and velocities at the throat and test sections.

(b) area of cross section of the test section.

(c) mass flow rate

(d) power required to drive the compressor.

[36.5 kPa, 258 k, 0.49 kg/m^3, 323 m/s, 528 m/s, 15.9 kg/s, 2182 kW.]

3.17 A supersonic wind tunnel nozzle is to be designed for Mach number 2 with a throat sectional area 0.1 m^2. The supply pressure and temperature at nozzle inlet where velocity is negligible are 70 kPa and 310 K respectively. Assume isentropic flow throughout the nozzle, determine mass flow rate, area at the test section, and the air properties at throat and test section. [$\dot{m}$ = 16 kg/s, 0.168 m^2, 0.089 bar, 172 K.]

4

DIABATIC FLOW

4.1 INTRODUCTION

Actual fluid flows are generally non-isentropic. An important example of non isentropic flow with heat transfer (diabatic flow). Flow involves adiabatic (no heat transfer) are generally nonisentropic also. In this section we consider the diabatic flow of an ideal gas through a constant cross section area duct with heat transfer. This type of flow is often referred to as Rayleigh flow. We also analyze the adiabatic flow of an ideal gas through a constant cross section area duct with friction.

This type of flow is often referred as Fanno flow. The concepts associated with Rayleigh and Fanno flows lead to further discussion of normal shock waves.

4.2. Constant area duct flow with heat transfer (diabatic flow)

The following assumptions are made for diabatic flow.

(1) Flow is steady

(2) Flow is one dimensional

(3) Flow is frictionless

(4) Flow takes place in a constant area section.

(5) The gas is perfect and hence has constant specific heats.

(6) No work transfer.

4.2.1 Rayleigh curve

For diabatic flow in a constant area duct without friction, the continuity and momentum equation, together with gas relations, give an equation whose graph is called the Rayleigh line.

The equation of confirmity is written as

$$\dot{m} = \rho V A$$

The mass velocity is written as

$$\frac{\dot{m}}{A} = G = \rho V \tag{4.1}$$

and the moment um equation is given by

$$p + \rho V^2 = constant \tag{4.2}$$

Combinig these two equations (4.1) and (4.2) we obtain.

$$p + \frac{G^2}{\rho} = constant \tag{4.3}$$

which is one of the Rayleigh line equations.

For a pure substance both p and ρ can be expressed as function of the enthalpy h and the entropy s so that, for given constants G and $p + \rho V^2$, equation (4.3) may be used to plot the variations of h with s. The resulting curve is known as a Rayleigh curve (Fig 4.1).

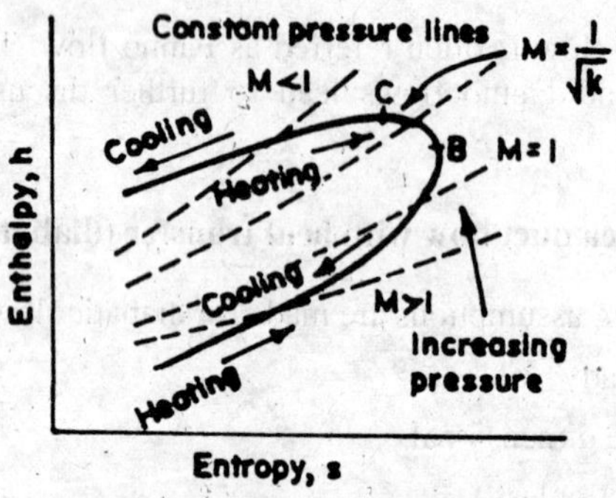

Fig 4.1 Rayleigh curve

Point in Fig 4.1 corresponds to sonic velocity and a Mach number of unity, which may be shown by differentiating equation (4.3) to obtain

$$dp - \frac{G^2}{\rho^2} d\rho = 0$$

or, $$\frac{dp}{d\rho} = \frac{G^2}{\rho^2} = \frac{\rho^2 V^2}{\rho^2} = V^2$$

or, $$a^2 = V^2$$ [From equation (2.3)]

or, $\dfrac{V^2}{a^2} = M^2$

or, M = 1

Therefore the flow cannot continue beyond M = 1 with either heat addition or heat rejection alone. However the combination of heating and cooling can allow a flow to cross M = 1 without discontiunity. At the point of maximum temperature (point C of Fig 4.1), the Mach number for an ideal gas is $1/\sqrt{k}$. The upper part of the Rayleigh curve corresponds to subsonic flow (M < 1), and the lower part to supersonic flow (M > 1). Regardless of the initial Mach number, with heat addition the flow state proceeds to the right and with heat rejection the flow state proceeds to the left along the Rayleigh line. Note from Fig 4.1 that for heating at points between the highest point on the Rayleigh line (point C) and point B, the enthaply of gas decrease. This means that heating in this region will cool the gas. Conversely, heat removal in this region between points B and C warm the gas.

4.2.2 Rayleigh flow relations

Fig 4.2 shows Rayleigh flow through a control volume. The flow parameters in a finite process at entry and exit of the control volume are also shown. The changes in the flow parameters occur due to the heat transfer Q.

The initial and final state points during the process shown for both subsonic and supersonic heating processes. During cooling the process will occur from state 2 to 1.

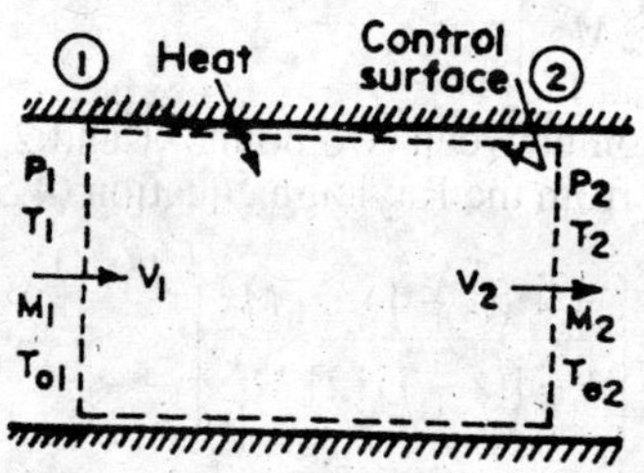

Fig 4.2 Rayleigh flow through a control volume.

4.2.3 Pressure ratio

An equation expressing the pressure ratio for a Rayleigh process is derived as follows :

The momentum equation is

$$p + \rho V^2 = \text{constant}$$

and for constant area section

$$p_2 + \rho_2 V_2^2 = p_1 + \rho_1 V_1^2$$

or, $$p_2 - p_1 + \rho_2 V_2^2 - \rho_1 V_1^2 = 0 \tag{4.4}$$

We know that

$$M^2 = \frac{V^2}{k\,RT} \text{ and } \frac{p}{\rho} = RT$$

Combining both the equation

$$M^2 = \frac{\rho\, V^2}{k\,p}$$

or, $$\rho\, V^2 = k\,PM^2$$

or, $$\rho_1\, V_1^2 = k\,p_1\, M_1^2 \text{ and } \rho_2\, V_2^2 = k\,p_2\, M_2^2$$

Substituting in equation (4.4)

$$p_2 - p_1 + k\,p_2\, M_2^2 - k\,p_2\, M_1^2 = 0$$

or, $$p_2\,(1 + k\,M_2^2) = p_1\,(1 + k\,M_1^2)$$

or, $$\frac{p_2}{p_1} = \frac{1 + k\,M_1^2}{1 + k\,M_2^2} \tag{4.5}$$

It was shown in equation (2.16) that the stagnation pressure, static pressure and Mach number are related by the following equation

$$\frac{p_o}{p} = \left(1 + \frac{k-1}{2}M^2\right)^{k/(k-1)}$$

Writing this equation between two points, 1 and 2, dividing one by the other and finally susbtituting from the Rayleigh equation (4.5), we get

$$\frac{p_{o2}}{p_{o1}} = \frac{1 + k\,M_1^2}{1 + k\,M_2^2}\left[\frac{1 + [(k-1)/2]\,M_2^2}{1 + [(k-1)/2]\,M_1^2}\right]^{k/(k-1)} \tag{4.6}$$

4.2.4 Temperature ratio

Using the continuity equation $\dot{m} = \rho_1 A_1 v_1 = \rho_2 A_2 v_2 =$ constant

For constant area flow

$$\rho_1/\rho_2 = \frac{V_2}{V_1} \tag{4.7}$$

and perfect gas relation

$$\frac{p_1}{\rho_1} = RT_1 \text{ and } \frac{p_2}{\rho_2} = RT_2$$

and division yields

$$\frac{T_2}{T_1} = \frac{p_2 \rho_1}{p_1 \rho_2} \tag{4.8}$$

By the definition of Mach number, we can write

$$M_1 = \frac{V_1}{\sqrt{k R T_1}} \text{ and } M_2 = \frac{V_2}{\sqrt{k R T_2}}$$

and division yields

$$\frac{V_2}{V_1} = \frac{M_2 \sqrt{T_2}}{M_1 \sqrt{T_1}} \tag{4.9}$$

Combining equation (4.7), (4.8), (4.9) and (4.5) we get

$$\frac{T_2}{T_1} = \left[\frac{M_2}{M_1}\left(\frac{1 + k M_1^2}{1 + k M_2^2}\right)\right]^2 \tag{4.10}$$

The upstream and downstream side, the appropriate temperature equations can be written as

$$\frac{T_{o1}}{T_1} = 1 + \frac{k-1}{2} M_1^2$$

and
$$\frac{T_{o2}}{T_2} = 1 + \frac{k-1}{2} M_2^2$$

Division then gives

$$\frac{T_{o2}}{T_{o1}} = \left(\frac{T_2}{T_1}\right)\left[\frac{1 + \left(\frac{k-1}{2}\right) M_1^2}{1 + \frac{k-1}{2} M_2^2}\right] \tag{4.11}$$

Combining equations (4.10) with equation (4.11), we get

$$\frac{T_{o2}}{T_{o1}} = \left[\frac{M_2}{M_1}\left(\frac{1 + k M_1^2}{1 + k M_2^2}\right)\right]^2 \left[\frac{1 + \left(\frac{k-1}{2}\right) M_1^2}{1 + \left(\frac{k-1}{2}\right) M_2^2}\right] \tag{4.12}$$

4.2.5 Working Rayleigh formula

The equation (4.12) is convenient because it is in terms of quantities which can be measured with relative ease. However, it is possible to modify it slightly and make it even more convenient for analytical work. Let $M_1 = 1$ and denote all properties at this point by the superscript asterisk. Thus $M_1 = M^* = 1$

$$T_1 = T^*, p_1 = p^* \text{ and } T_{o_1} = T_o^*$$

The condition yields two commonly used expressions namely,

$$\frac{T}{T^*} = \frac{M^2 (1+k)^2}{(1+k M^2)^2} \tag{4.13}$$

and
$$\frac{T_o}{T_o^*} = \frac{2 M^2 (k+1)\left[1+\frac{(k-1)}{2} M^2\right]}{(1+k M^2)^2} \tag{4.14}$$

Similarly, we obtain the following formula

$$\frac{V}{V^*} = \frac{\rho^*}{\rho} = \frac{(k+1) M^2}{1+k M^2} \tag{4.15}$$

$$\frac{p}{p^*} = \frac{k+1}{1+k M^2} \tag{4.16}$$

$$\frac{p_o}{p_o^*} = \frac{k+1}{1+k M^2}\left[\frac{2\left(1+\frac{k-1}{2} M^2\right)}{k+1}\right]^{k/k-1} \tag{4.17}$$

Using these normalised expressions, the ratios of properties at two sections where Mach numbers are M_1 and M_2 are given by expressions of the type

$$\frac{T_{o2}}{T_{o1}} = \frac{(T_o/T_o^*)M_2}{(T_o/T_o^*)\, M_1} \tag{4.18}$$

Where $(T_o/T_o^*)_{M_2}$ is found form the equation (4.14), and so forth.

4.2.6 Working tables and charts

Tabulated values of $\frac{T_o}{T_o^*}$, $\frac{T}{T^*}$, $\frac{p}{p^*}$, $\frac{p_o}{p_o^*}$, $\frac{\rho^*}{\rho}$ and $\frac{V}{V^*}$ as a function of Mach number will be found in the appropriate tables in any book of compressible flow. For each value of k different tables must be constructed.

The various property ratios as obtained from the Rayleigh line table have been plotted in Fig 4.3. It should be noted at once that the curves in the figure 4.3 do not denote paths or process in the usual sense. Instead, each curve in Fig 4.3 represents the locus of the particular property ratio as defined by the Mach number.

In Fig 4.3, the curve of $\frac{T_o}{T_o^*}$ is probably the most basic because the direction of the stagnation temperature change indicates the direction of heat transfer.

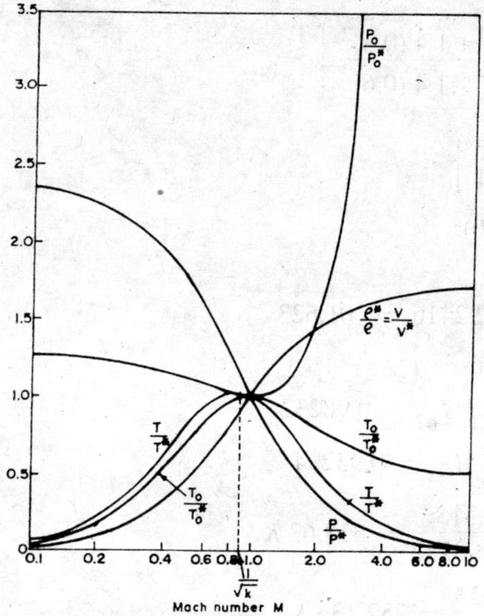

Fig 4.3 Rayleigh flow through a control volume.

TABLE 4.1 Variation of flow parameters

Property	Heating		Cooling	
	M > 1	M < 1	M > 1	M < 1
T_o	Increases	Increases	Decreases	Decreases
p	Increases	D	D	I
p_o	Decreases	D	I	I
V	Decreases	I	I	D
T	Increases always	I when M < 0.85 D when M > 0.85	D always	D when M < 0.85 I when M > 0.85

D for decrease, I for Increase.

The variation of the flow properties for heating and cooling in each domain are shown in table 4.1.

ILLUSTRATIVE EXAMPLES

4.1 Air flows through a short tube without friction. Heat is supplied to increase the initial Mach number of 0.3 at a temperature of 313 K to a final Mach number of 0.6. How much heat must be supplied per unit mass of air ?

Solution We know that

$$\frac{T_2}{T_1} = \frac{M_2^2}{M_1^2}\left(\frac{1 + k\,M_1^2}{1 + k\,M_2^2}\right)^2$$

$$= \left(\frac{0.6}{0.3}\right)^2 \left[\frac{1 + 1.4\,(0.3)^2}{1 + 1.4\,(0.6)^2}\right]^2$$

$$= 4 \times \left(\frac{1.126}{1.504}\right)^2$$

$$= 4 \times (0.7486)^2$$

$$= 2.2416$$

$$T_2 = 313 \times 2.2416 = 701.623$$

From isentropic tables

At $M_1 = 0.3,\ T_1/T_{o1} = 0.98232$

and at $M_2 = 0.6,\ T_2/T_{o2} = 0.93284$

Therefore, $T_{o1} = \dfrac{313}{0.98232} = 318.63\ K$

and $T_{o2} = \dfrac{701.623}{0.93284} = 752.136\ K$

Thus $q = c_p\,(T_{o2} - T_{o1}) = 1.005\,(752.136 - 318.63)$

$= 1.005 \times 433.506$

$= 435.674\ kJ/kg$ Ans

4.2 As heat is added to a subsonic frictionless gas flowing in a constant area duct, the Mach number, increases to a maximum of 1 before choking occurs. The gas temperature increases to a maximum at a Mach number of $1/\sqrt{k}$, then decreases to T^*, the value associated with a Mach number of 1.

(a) At what subsonic Mach number is the gas temperature T = T^* for air ?

(b) What is T_{max}/T^* for air ?

Solution

(a) We know that

$$\frac{T_2}{T_1} = \frac{M_2^2}{M_1^2}\left(\frac{1 + k\,M_1^2}{1 + k\,M_2^2}\right)^2$$

At $\quad T_1 = T^*$, then

$$1 = \frac{1}{M_1^2}\frac{(1+k\,M_1^2)^2}{(1+k)^2}$$

$$M_1^2\,(1+1.4)^2 = (1+1.4\,M_1^2)^2$$

$$5.76\,M_1^2 = 1 + 2\times 1.4\,M_1^2 + 1.96\,M_1^4$$

or, $\quad 5.76\,M_1^2 = 1 + 2.8\,M_1^2 + 1.96\,M_1^4$

$$1.96\,M_1^4 - 2.96\,M_1^2 + 1 = 0$$

$$M_1^4 - 1.51\,M_1^2 + 0.51 = 0$$

Solving the above equation, we get the value of $M_1 = 0.714$

(b) We know that

$$\frac{T_2}{T_1} = \frac{M_2^2}{M_1^2}\frac{\left(1+k\,M_1^2\right)^2}{\left(1+M_2^2\right)^2}$$

$$\frac{T_{max}}{T^*} = \frac{1/k}{1}\frac{(1+k)^2}{(1+1)^2}$$

$$\frac{T_{max}}{T^*} = \frac{(1+K)^2}{4K}\frac{(1+1.4)^2}{4\times .4} = \frac{(2.4)^2}{5.6}$$

$$= \frac{5.76}{5.6} = 1.02857$$ Ans

4.3 Air flows with negligible friction in a constant area duct. At section one, the flow properties are $t_1 = 60.4°C$, $p_1 = 135$ kPa absolute and velocity 732 m/s. Heat is added to the flow between section one and section two, where the Mach number is 1.2. Determine the flow properties at section two, the heat transfer per unit mass and the entropy change.

Solution To obtain property ratios from tables, we need both Mach numbers,

$$a_1 = \sqrt{k\,RT_1} = \sqrt{1.4\times 287\times 333.4} = 366\ m/s$$

$$M_1 = \frac{V_1}{a_1} = \frac{732}{366} = 2$$

From Rayleigh line flow function

M	T_o/T_o^*	p_o/p_o^*	T/T^*	p/p^*	V/V^*
2	0.7934	1.503	0.5289	0.3636	1.455
1.2	0.9787	1.019	0.9119	0.7958	1.146

The critical properties are constant and using the abstract of the isentropic tables given above.

$$\frac{p_2}{p_1} = \frac{p_2/p^*}{p_1/p^*} = \frac{0.7958}{0.3636} \approx 2.19$$

$$\therefore \quad p_2 = 2.19\, p_1 = 2.19 \times 135 = 295.65\ kPa$$

$$\frac{T_2}{T_1} = \frac{T_2/T^*}{T_1/T^*} = \frac{0.9119}{0.5289} \approx 1.72$$

$$\therefore \quad T_2 = 1.72\, T_1 = 1.72 \times 333.4 = 573.448\ K$$ Ans

$$\frac{V_2}{V_1} = \frac{V_2/V^*}{V_1/V^*} = \frac{1.146}{1.455} = 0.7876$$

$$\therefore \quad V_2 = 0.7876\, V_1 = 0.7876 \times 732 = 576.54\ m/s$$ Ans

$$p_2 = \rho_2 R T_2$$

$$295.65 = \rho_2 \times 0.287 \times 573.448$$

$$\rho_2 = 1.796\ kg/m^3$$ Ans

From the isentropic flow tables

At $M_1 = 2$

$$\frac{T}{T_o} = \frac{T_1}{T_{o1}} = 0.5556$$

$$\therefore \quad T_{o1} = \frac{T_1}{0.5556} = \frac{333.4}{0.5556} = 600\ K$$

and at $M_2 = 1.2$

$$\frac{T}{T_o} = \frac{T_2}{T_{o2}} = 0.7764$$

$$\therefore \quad T_{o2} = \frac{T_2}{0.7764} = \frac{573.448}{0.7764} = 738.598\ K$$ Ans

Therefore, $q = h_{o_2} - h_{o_1} = cp\,(T_{o2} - T_{o_1})$

$= 1.005\ (738.598 - 600) = 139.29$ kJ/kg Ans

$$s_2 - s_1 = Cp \ln \frac{T_2}{T_1} - R \ln p_2/p_1$$

$$= 1.005 \ln \frac{573.448}{333.4} - 0.287 \ln \frac{295.65}{135}$$

$$= 0.545 - 0.2249$$

$$= 0.32 \text{ kJ/kg K.}$$

Check From isentropic tables

At $M_1 = 2$,

$$\frac{p}{p_o} = \frac{p_1}{p_{o1}} = 0.1278$$

$$\therefore \quad p_{o1} = \frac{p_1}{0.1278} = \frac{135}{0.1278} = 1056.338\ kPa$$

and at $M_2 = 1.2$,

$$\frac{p}{p_o} = \frac{p_2}{p_{o2}} = 0.4124$$

$$\therefore \quad p_{o2} = \frac{p_2}{0.4124} = \frac{196}{0.4124} = 475.26\ kPa$$

Thus, $p_{o1} > p_{o2}$ as expected for heating process.

4.4 Air flows steadily and without friction at the rate of 1.83 kg/s through a duct with cross sectional area of 0.02 m^2. At the duct inlet, the temperature and absolute pressure are 260°C and 126 kPa, respectively. The exit flow discharges subsconically to atmospheric pressure. Determine the Mach number, temperature, and stagnation temperature at the duct outlet, and the heat transfer rate.

Solution

$$p_1 = \rho_1 R T_1$$

$$126 = \rho_1 \times 0.287 \times 533$$

$$\rho_1 = 0.82368\ kg/m^3$$

$$\dot{m} = \rho_1 A V_1$$

$$1.83 = 0.82368 \times 0.02 \times V_1$$

$$V_1 = 111.086\ m/s$$

$$a_1 = \sqrt{k R T_1}$$

$$= \sqrt{1.4 \times 287 \times 533} = 462.77\ m/s$$

$$M_1 = \frac{V_1}{a_1} = \frac{111.086}{462.77} = 0.24$$

From isentropic tables

At $M_1 = 0.24$

$$\frac{T_1}{T_{o1}} = 0.988 \therefore T_{o1} = \frac{533}{0.988} = 539.4736\ K$$

From Rayleigh flow

At $M_1 = 0.24$

$$\frac{T_{o_1}}{T_o^*} = 0.2395\ ,\ \frac{T_1}{T^*} = 0.284\ ,\ \frac{p_1}{p^*} = 2.221$$

$$T_o^* = \frac{539.4736}{0.2395} = 2252.5\ K$$

$$T^* = \frac{533}{0.284} = 1876\ K$$

$$p^* = \frac{126}{2.221} = 56.73\ k\ Pa$$

The atmospheric pressure $p_2 = 101$ kPa.

At section two,

$$\frac{p_2}{p^*} = \frac{101}{56.73} = 1.78$$

From Rayleigh flow

At $\frac{p_2}{p^*} = 1.78\ ,\ M_2 \approx 0.5$ **Ans**

$$\frac{T_2}{T^*} = 0.79\ ,\ \frac{T_{o2}}{T_o^*} = 0.691$$

$$T_2 = 1876 \times 0.79 = 1482.04\ K$$

$$T_{o2} = 2252.5 \times 0.691 = 1556.47\ K$$

$$\dot{Q} = \dot{m}\ \ c_p\,(T_{o2} - T_{o1})$$

$$= 1.83 \times 1.005\,(1556.47 - 539.4736) = 1870.42\ kJ/s.$$ **Ans**

4.5 Air flows without friction through a duct of 75 mm diameter. The inlet Mach number, stagnation temperature, and absolute pressure are $M_1 = 0.2$, $T_o = 278$ K, and $p_1 = 273$ kPa, respectively. Heat is added at a rate equivalent to 219 kJ/kg of flowing air. Determine the stagnation temperature and Mach number at the duct outlet, and the changes in entropy and stagnation pressure for the process.

Solution

From isentropic tables

At $M_1 = 0.2, p_1/p_{o1} = 0.9725$

$$\text{Therefore, } p_{o1} = \frac{273}{0.9725} = 280.72\ kPa$$

$$q = c_p (T_{o2} - T_{o1})$$

$$219 = 1.005 (T_{o2} - 278)$$

$$T_{o2} = 495.91\ K$$ Ans

From Rayleigh tables.

$$\frac{T_{o1}}{T_o^*} = 0.167$$

and $$\frac{p_{o1}}{p_o^*} = 1.224$$

$$\text{Therefore, } p_o^* = \frac{280.72}{1.224} = 229.35\ kPa$$

Then

$$\frac{T_{o2}}{T_o^*} = \frac{T_{o2}}{T_{o1}} \times \frac{T_{o1}}{T_o^*}$$

$$= \frac{495.91}{278} \times 0.167$$

$$= 1.7838 \times 0.167$$

$$= 0.2979$$

From Rayleigh tables,

At $\frac{T_{o2}}{T_o^*} = 0.2979,\ M_2 \approx 0.28$ Ans

$$\frac{p_{o2}}{p_o^*} = 1.198$$

$$\text{Therefore, } p_{o2} = 1.198 \times 229.35 = 274.757\ k\,Pa$$

$$p_{o1} - p_{o2} = 280.72 - 274.757 = 5.963\ k\,Pa$$ Ans

$$s_2 - s_1 = Cp \ln \frac{T_{o2}}{T_{o1}} - R \ln \frac{p_{o2}}{p_{o1}}$$

$$= 1.005 \ln \frac{495.91}{278} - 0.287 \ln \frac{274.757}{280.72}$$

$$= 0.58166 + 6.162 \times 10^{-3}$$

$$= 0.587829 \text{ kJ/kg K.}$$ Ans

4.6 Air flow in constant area duct without friction at an initial temperature of 40°C and an initial Mach number of 0.5. Heat is added to the air at a rate of 56 kJ/kg between section one and two.

(a) What is the stagnation temperature at each section ?

(b) What is the ratio of stagnation temperature at each section to that where the Mach number is unity ? (Find T_{o1}/T_o^* and T_{o2}/T_o^*.)

(c) What is the Mach number at section two ?

Solution

(a) From isentropic tables

At $M_1 = 0.5$, $\frac{T_1}{T_{o1}} = 0.95238$

Therefore, $T_{o1} = \frac{T_1}{0.9538} = \frac{313}{0.95238} = 328.65\ K$

$$q = c_p (T_{o2} - T_{o1})$$

$$56 = 1.005 (T_{o2} - 328.65)$$

$$T_{o2} = 55.72 + 328.65 = 384.37\ K$$ Ans

(b) Using Rayleigh tables

At $M_1 = 0.5$, $\frac{T_{o1}}{T_o^*} = 0.691$

$$\frac{T_{o2}}{T_o^*} = \frac{T_{o2}}{T_{o1}} \times \frac{T_{o1}}{T_o^*}$$

$$= \frac{384.37}{328.65} \times 0.691$$

$$= 1.1695 \times 0.691 = 0.808$$ Ans

(c) Using Rayleigh tables,

At $\quad \frac{T_{o2}}{T_o^*} = 0.808$

$M_2 = 0.59$ Ans

4.7 In the frictionless flow of air through a 100 mm diameter duct, 1.42 kg/s enters at a temperature of 52°C and an absolute pressure of 60 kPa. Determine the amount of heat that must be added to choke the flow, and the fluid properties at the choked state.

Solution

$$A = \frac{\pi}{4} D^2 = \frac{\pi}{4} (0.1)^2 = 7.85 \times 10^{-3}$$

$$p_1 = \rho_1 RT_1$$

$$60 = \rho_1 \times 0.287 \times 325$$

$$\rho_1 = 0.643259\ kg/m^3$$

$$\dot{m} = \rho_1 A V_1$$

$$1.42 = 0.643259 \times 7.85 \times 10^{-3} \times V_1$$

$$V_1 = 281.2\ m/s$$

$$a_1 = \sqrt{k RT_1} = \sqrt{1.4 \times 287 \times 325} = 361.365$$

$$\therefore \quad M_1 = \frac{V_1}{a_1} = \frac{281.2}{361.365} = 0.778 \approx 0.78$$

From isentropic tables

At $\quad M_1 = 0.78, \frac{T_1}{T_{o1}} = 0.892$

$$\frac{p_1}{p_{o1}} = 0.669, \; T_{o1} = \frac{325}{0.892} = 364.35\ K$$

$$p_{o1} = \frac{p_1}{0.669} = \frac{60}{0.669} = 89.686\ k\,Pa$$

From Rayleigh tables

At $\quad M_1 = 0.78\,, \frac{p_1}{p^*} = 1.296\,, \frac{\rho^*}{\rho} = 0.788 = \frac{V}{V^*}$

$$\frac{T_{o1}}{T_o^*} = 0.955\,, \frac{p_{o1}}{p_o^*} = 1.023\,, \frac{T_1}{T^*} = 1.022$$

The flow is choked.

At $M_2 = 1, \dfrac{T_{o_2}}{T_{o*}} = 1, \dfrac{p_{o2}}{p_o^*} = 1, \dfrac{T_2}{T^*} = 1$ ----

Therefore

$$\frac{T_{o2}}{T_o^*} = \frac{T_{o2}}{T_{o1}} \times \frac{T_{o1}}{T_o^*}$$

$$1 = \frac{T_{o2}}{364.35} \times 0.955$$

$$\therefore \quad T_{o2} = 381.518\ K \qquad \text{Ans}$$

Similarly,

$$p_{o2} = 89.686/1.023 = 87.67\ kPa \qquad \text{Ans}$$

$$T_2 = 325/1.022 = 318\ K$$

$$p_2 = 60/1.296 = 46.296\ kPa$$

$$\rho_2 = 0.643 \times 0.788 = 0.506\ kg/m^3$$

$$V_2 = \frac{281.2}{0.788} = 356.85\ m/s$$

$$q = c_p\,(T_{o2} - T_o1)$$

$$= 1.005\,(381.518 - 364.35)$$

$$= 17.25 \text{ kJ/kg} \qquad \text{Ans}$$

4.8 A converging-diverging nozzle feeds a short duct of constant area. At the duct inlet, the local isentropic stagnation conditions are $T_{o1} = 700\ K, p_{o1} = 800\ kPa$ and the Mach number is $M_1 = 3$. A short distance down the duct at section two, the density is $\rho_2 = 0.334\ kg/m^3$. Assuming frictionless flow in the duct, determine the speed, pressure, and Mach number at section two and the heat transfer between the inlet and section two.

Solution

From isentropic tables

At $M_1 = 3, \dfrac{T_1}{T_{o1}} = 0.357, \dfrac{p_1}{p_{o1}} = 0.0272$

$$\therefore \quad T_1 = 700 \times 0.357 = 249.9\ K$$

$$p_1 = 800 \times 0.0272 = 21.76\ k\,Pa$$

From Rayleigh tables

At $M_1 = 3, \frac{T_{o1}}{T_o^*} = 0.603, \frac{T_1}{T^*} = 0.295, \frac{p_1}{p^*} = 0.181$

$$\frac{V_1}{V^*} = \frac{\rho^*}{\rho_1} = 1.63$$

$$\therefore \quad T_o^* = \frac{700}{0.603} = 1160.86\ K$$

$$T^* = \frac{249.9}{0.295} = 847.1\ K$$

$$p^* = \frac{21.76}{0.181} = 120.22\ k\,Pa$$

$$p_1 = \rho_1 R T_1$$

$$21.76 = \rho_1 \times 0.287 \times 249.9$$

$$\rho_1 = 0.3034\ kg/m^3$$

$$\rho^* = 1.63 \times 0.3034 = 0.4945\ kg/m^3$$

At section 2, $\frac{\rho^*}{\rho_2} = \frac{0.4945}{0.334} = 1.48$

From Rayleigh tables, $M_2 = 1.98$ Ans

$$\frac{T_{o2}}{T_o^*} = 0.771, \frac{T_2}{T^*} = 0.557, \frac{p_2}{p^*} = 0.377$$

$$T_{o2} = 1160.86 \times 0.771 = 895.02\ K$$

$$T_2 = 847.1 \times 0.557 = 471.83\ K$$

$$p_2 = 120.22 \times 0.377 = 45.32\ kPa$$ Ans

$$a_2 = \sqrt{k\,RT_2} = \sqrt{1.4 \times 287 \times 471.83} = 435.41$$

$$V_2 = M_2\, a_2 = 1.98 \times 435.41 = 862.11\ m/s$$ Ans

$$q = c_p\ (T_{o2} - T_o1)$$

$$= 1.005\ (862.11 - 700) = 162.92\ kJ/kg$$ Ans

4.9 Air is drawn from the standard atmospheric condition through a smooth converging nozzle into the inlet manifold of a racing engine. Flow is isentropic through the nozzle. Air leaves the nozzle and enters the manifold at Mach number M = 0.3. In the manifold, gasoline is evaporated in a short section of constant area duct where friction is negligible. The effect is toremove 60.3 kJ/kg of thermal energy from the air stream. The fluid properties are those of pure air. Find (a) the Mach number, and (b) the pressure at the outlet of the evaporation section. Take the standard atmospheric values of pressure and temperature 101 kPa and 288 K.

Solution

Flow through the nozzle is isentropic.

From isentropic tables

At $M_1 = 0.3$

$$\frac{T_1}{T_o} = 0.9823 \text{ and } \frac{p_1}{p_{o1}} = 0.9395$$

Therefore, $T_1 = 288 \times 0.9823 = 282.9\ K$

and $p_1 = 101 \times 0.9395 = 94.8895\ k\,Pa$

From Rayleigh tables

At $M_1 = 0.3,\ T_{o1}/T_o^* = 0.347,\ \frac{p_1}{p^*} = 2.131$

Therefore,

$$T_o^* = \frac{T_{o1}}{0.347} = \frac{288}{0.347} = 829.97\ K$$

$$p^* = \frac{p_1}{2.131} = \frac{94.8895}{2.131} = 44.528\ k\,Pa$$

$$Q = c_p\,(T_{o2} - T_{o1})$$

$$-60.3 = 1.005\,(T_{o2} - 288)$$

$$T_{o2} = 228\ K$$

Therefore, $\frac{T_{o2}}{T_o^*} = \frac{228}{829.97} = 0.2747$

From Rayleigh tables

At $\frac{T_{o2}}{T_o^*} = 0.2747,\ M_2 = 0.26$ Ans

$$\frac{p_2}{p^*} = 2.193$$

$$p_2 = p^* \times 2.193 = 44.528 \times 2.193 = 97.65\ kPa$$ Ans

4.10 A fuel-air mixture with the thermodynamic properties of pure air enters a combustor of constant area. The stagnation temperature of the inlet stream is constant at 400 k. Friction is negligence. When heat is added to the flow at the rate of 1.4 MJ/kg, flow at the duct exit is choked. At this condition the static pressure at the inlet is 144.6 kPa and the pressure at the duct exit is 64.7 kPa. Calculate

(a) the combustor outlet temperature.

(b) the combustor inlet Mach number, and

(c) the loss in stagnation pressure through the combustor.

Solution

The flow is choked means that the exit Mach number $M_2 = 1$.

From isentropic tables

At $M_2 = 1$

$$\frac{T_2}{T_{o2}} = 0.834, \quad \frac{p_2}{p_{o2}} = 0.5283$$

$$Q = m\, c_p\, (T_{o_2} - T_{o1})$$

$$1.4 \times 1000 = 1 \times 1.005\,(T_{o2} - 400)$$

$$T_{o2} = 1793\ K$$

Therefore, $T_2 = 1793 \times 0.834 = 1495.4\ K$ Ans

$$p_{o2} = \frac{p_2}{0.5283} = \frac{64.7}{0.5283} = 122.468\ kPa$$

From Rayleigh tables, At $M_2 = 1$.

$$p_2/p^* = 1$$

Therefore, $p^* = 64.7\ k\,Pa$

and $\dfrac{p_1}{p^*} = \dfrac{144.6}{64.7} = 2.2349$

From Rayleigh tables

At $\dfrac{p_1}{p^*} = 2.2349$, $M_1 \approx 0.23$ Ans

From isentropic tables

At $M_1 = 0.23$

$$\frac{p_1}{p_{o1}} = 0.9638$$

Therefore, $p_{o1} = \frac{p_1}{0.9638} = \frac{144.6}{0.9638} = 150.031\ kPa$

Therefore, loss in stagnation pressure

$$\left(p_{o2} - p_{o1}\right) = 122.468 - 150.031 = -27.563\ kPa$$ Ans

4.11 Air from an aircraft inlet system enters the engine combustion chamber where heat is added during a frictionless process in a tube with constant area of 0.01 m^2. The local isentropic stagnation temperature and Mach number entering the combustor are 427 K and 0.3. The mass flow rate is 0.5 kg/s. When the rate of heat transfer is set of 404 kW, the flow leaves the combustor at 1026 k and 22.9 kPa. Determine for this process (a) the Mach number at combustor outlet, (b) the static pressure at the combustor inlet, and (c) the change in local isentropic stagnation pressure during the heat addition process.

Solution

$$\dot{Q} = \dot{m}\ c_p (T_{o2} - T_{o1})$$

$$404 = 0.5 \times 1.005\ (T_{o2} - 427)$$

$$T_{o2} = 803.98 + 427 = 1230.98\ K$$

From Rayleigh tables

At $M_1 = 0.3, \frac{T_{o1}}{T_o^*} = 0.347$

Therefore, $T_o^* = \frac{427}{0.347} = 1230.54\ K$

Since $T_{o2} \approx T_o^*$, then $M_2 = 1$ Ans

and $p_2 = p^* = 22.9\ kPa$

From Rayleigh tables

At $M_1 = 0.3, \frac{p_1}{p^*} = 2.131,$

$\therefore \quad p_1 = 2.131 \times 22.9 = 48.8\ kPa$

From isentropic tables

At $M_1 = 0.3, \dfrac{p_1}{p_{o1}} = 0.939,$

$$p_{o1} = \frac{48.8}{0.939} = 51.97\ kPa$$

From isentropic table

At $M_2 = 1, \dfrac{p_2}{p_{o_2}} = 0.528,$

$$\text{Therefore, } p_{o2} = \frac{p_2}{0.528} = \frac{22.9}{0.528} = 43.37\ kPa$$

Therefore change in local stagnation pressure

$$= p_{o2} - p_{o1} = \Delta p_o$$

$$\Delta p_o = 43.37 - 51.97 = -8.598\ kPa \qquad \text{Ans}$$

and $\dfrac{\Delta p_o}{p_{o1}} = -\dfrac{8.598}{51.97} = -0.165 = -16.5\ \%$ Ans

4.12 Air enters the inlet manifold of a S.I. engine at the rate of 0.308 kg/s with stagnation temperature $T_{o1} = 300\ K$. At a certain location, $M_1 = 0.3$, and $p_1 = 94$ kPa. The air stream is cooled by the evaporation of alcohol fuel. After the evaporation section, $M_2 = 0.2$. The effects of friction and area change are negligible. Treat the fluid as pure air. Find the flow area. Calculate the heat transferred per unit mass of air, the change in stagnation pressure for the process, and the entropy change.

Solution

From isentropic tables

At $M_1 = 0.3, \dfrac{T_1}{T_{o1}} = 0.9823, \dfrac{p_1}{p_{o1}} = 0.9395$

Therefore, $T_1 = 300 \times 0.9823 = 294.69\ K.$

and $p_{o1} = 94/0.9395 = 100.05\ kPa.$

$$p_1 = \rho_1 RT_1$$

$$94 = \rho_1 \times 0.287 \times 294.69$$

$$\therefore \quad \rho_1 = 1.11\ kg/m^3$$

$$a_1 = \sqrt{k\,RT_1}$$

$$= \sqrt{1.4 \times 287 \times 294.69}$$

$$= 344.1 \text{ m/s}$$

$$V_1 = M_1 a_1 = 0.3 \times 344.1 = 103.23\ m/s$$

$$\dot{m} = \rho_1 A V_1$$

$$0.308 = 1.11 \times A \times 103.23$$

$$A = 2.68 \times 10^{-3}\ m^2$$ Ans

From Rayleigh tables

At $M_1 = 0.3, \dfrac{T_{o1}}{T_o^*} = 0.347, \dfrac{p_{o1}}{p_o^*} = 1.198$

Therefore, $T_o^* = \dfrac{300}{0.347} = 864.55\ K$

$$p_o^* = \frac{100.053}{1.198} = 83.51\ kPa$$

At $M_1 = 0.28, \dfrac{T_{o2}}{T_o^*} = 0.31, \dfrac{p_{o2}}{p_o^*} = 1.206$

Therefore, $T_{o2} = 0.31 \times 864.55 = 268\ K$

$$p_o 2 = 1.206 \times 83.51 = 100.71\ kPa$$

$$\therefore \quad p_{o2} - p_{o1} = 100.71 - 100.053 = 0.657\ kPa$$

$$Q = c_p (T_{o2} - T_{o1})$$

$$= 1.005\,(268 - 294.69) = -26.82\ kJ/kg$$ Ans

$$s_2 - s_1 = c_p \ln \frac{T_{o2}}{T_{o1}} - R \ln \frac{p_{o2}}{p_{o1}}$$

$$= 1.005 \ln \frac{268}{294.69} - 0.287 \ln \frac{100.71}{100.053}$$

$$= -0.0954 - 1.8784 \times 10^{-3}$$

$$= -0.09729\ kJ/kg\ K \pm -97.29\ J/kg\ K$$ Ans

4.13 The condition of a gas in a combustor at entry are p_1 = 34.3 kPa, t_1 = 37°C and V_1 = 60 m/s. Determine the Mach number pressure, temperature and velocity at the exit if the increase in stagnation enthalpy of the gas between entry and exit is 1.1725 MJ/kg.

Solution

$$a_1 = \sqrt{k\,RT_1} = \sqrt{1.4 \times 287 \times 310} = 352.97\ m/s$$

$$M_1 = \frac{V_1}{a_1} = \frac{60}{352.97} = 0.17$$ Ans

From isentropic tables

$$\text{At } M_1 = 0.17, \frac{T_1}{T_{o1}} = 0.9943$$

$$\text{Therefore, } T_{o1} = \frac{T_1}{0.9943} = \frac{310}{0.9943} = 311.77\ K$$

From Rayleigh tables

$$\text{At} \quad M_1 = 0.17, \frac{p_1}{p^*} = 2.306, \frac{T_1}{T^*} = 0.154,$$

$$\frac{T_{o1}}{T_o^*} = 0.129,\ V/V^* = \frac{\rho^*}{\rho} = 0.0665$$

$$\text{Therefore, } T_o^* = \frac{T_{o1}}{0.129} = \frac{311.77}{0.129} = 2416.87$$

$$Q = \Delta h_o = c_p (T_{o2} - T_{o1})$$

$$1172.5 = 1.005 (T_{o2} - T_{o1})$$

$$T_{o2} = 1166.66 + 311.77 = 1478.436$$

$$\text{Therefore, } T_{o2}/T_o^* = \frac{1478.436}{2116.87} = 0.6117$$

From Rayleigh tables

$$\text{At} \quad T_{o2}/T_o^* = 0.6117, M_2 = 0.45$$ Ans

$$\frac{p_2}{p^*} = 1.87, \frac{T_2}{T^*} = 0.7075, \frac{V}{V^*} = 0.378$$

$$T_2 = \frac{0.7075}{0.154} \times 310 = 1424.18\ K$$ Ans

$$p_2 = \frac{1.87}{2.306} \times 34.3 = 27.8148\ kPa$$ Ans

$$V_2 = \frac{0.378}{0.0665} \times 60 = 341.05\ m/s$$ Ans

4.14 A jet aircraft cruises at Mach number 0.85 at 10 km altitude on a standard day. Air is slowed to Mach number 0.3 relative to the aircraft by the inlet system. Flow in the inlet system is adiabatic but friction causes the absolute stagnation pressure to drop by 5 percent. Air from the inlet is compressed to $p_2 = 30\ p_1$ in the engine's compressor section. Flow in the compressor is adiabatic, but the stagnation temperature of the stream increased by 750 K due to work addition. The Mach number of the flow leaving the compressor and entering the combustor is 0.4. Heat is added in the combustor, where the maximum temperature must be held to 1630 K or less. Evaluate for these conditions (a) the maximum possible heat addition in the combustor; (b) the combustor exit Mach number for part (a), and (c) the change in stagnation pressure across the combustor.

Solution At the inlet at 10 km altitude.

$$M_a = 0.85, T_a = 223\ k \text{ and } p_a = 26.5\ kPa$$

From isentropic table

At $M_a = 0.85, \frac{T_a}{T_{oa}} = 0.874, \frac{p_a}{p_{oa}} = 0.624$

Therefore, $T_{oa} = \frac{223}{0.874} = 255.15\ K$

$$p_{oa} = \frac{26.5}{0.624} = 42.468\ kPa$$

At section (1), $T_{o1} = T_{oa} = 255.15\ K$

$$p_{o1} = 0.95\ p_{oa} = 0.95 \times 42.468 = 40.3446\ kPa$$

From isentropic tables

At $M_1 = 0.3, \frac{T_1}{T_{o1}} = 0.982, \frac{p_1}{p_{o1}} = 0.939$

Therefore, $T_1 = 0.982 \times 255.15 = 250.55\ K$

$$p_1 = 0.939 \times 40.3446 = 37.88\ kPa$$

At section (2), $p_2 = 30\ p_1 = 30 \times 37.88 = 1136.5\ kPa$

$$T_{o2} = T_{o1} + 750 = 255.15 + 750 = 1005.15\ K$$

From isentropic tables

At $M_2 = 0.4, \frac{T_2}{T_{o2}} = 0.969, \frac{p_2}{p_{o2}} = 0.895$

Therefore, $T_2 = 0.969 \times 1005.15 = 973.99\ K$

$$p_{o2} = \frac{1136.5}{0.895} = 1269.832\ kPa$$

For maximum heat adition, $M_3 = 1$

From Rayleigh flow check for temperature

At $M_2 = 0.4, \dfrac{T_{o2}}{T_o^*} = 0.516, \dfrac{p_{o2}}{p_o^*} = 1.152$

$$T_o^* = \frac{1005.15}{0.516} = 1947.965 = T_3$$

$$p_o^* = \frac{1269.832}{1.152} = 1102.28\ kPa$$

$$= p_{o3}$$

From isentropic table

At $M_3 = 1 \dfrac{T_3}{T_{o3}} = 0.834$

Therefore, $T_3 = 0.834 \times 1947.965 = 1624.6\ K$

$$= T^*$$

However for Rayleigh line flow $T = T_{max}$

When $M = 1/\sqrt{k} = 0.845$

From Rayleigh table

At $M = 0.845, \dfrac{T}{T^*} = 1.029$

Therefore, $T = 1.029 \times 1624.6 = 1671.7$ K

Thus with $M_3 = 1$, maximum temperature in combustor is okay.

Change in stagnation pressure

$$\left(p_{o3} - p_{o2}\right) = 1102.28 - 1269.832 = -167.552\ kPa \qquad \text{Ans}$$

$$Q = c_p (T_{o3} - T_{o2})$$

$$= 1.005\,(1947.965 - 1005.15) = 947.529\ kJ/kg \qquad \text{Ans}$$

4.15 A combusion chamber in a gas turbine plant receives air at 77°C, 55 kPa and 75 m/s. The air fuel ratio is 29 and the calorific value of the fuel is 41870 kJ/kg for the gas determine;

(a) the initial and final Mach numbers,

(b) final pressure, temperature and velocity of the gas,

(c) percent stagnation pressure loss in the combusion chamber, and

(d) the maximum stagnation temperature attainable. Take k = 1.4 and R = 287 J/kg K.

Solution

(a) $a_1 = \sqrt{1.4 \times 287 \times 350} = 375\ m/s$

$$M_1 = \frac{V_1}{a_1} = \frac{75}{375} = 0.2$$ Ans

From isentropic tables

At $M_1 = 0.2, \frac{p_1}{p_{o1}} = 0.973, \frac{T_1}{T_{o1}} = 0.992$

$$p_{o1} = \frac{55}{0.973} = 56.526\ k\ Pa$$

$$T_{o1} = \frac{350}{0.992} = 352.82\ k$$

From Rayleigh tables

At $M_1 = 0.2,$

$$\frac{p_1}{p^*} = 2.273, \frac{p_{o1}}{p_o^*} = 1.235, \frac{T_1}{T^*} = 0.207$$

$$\frac{T_{o1}}{T_o^*} = 0.174 \text{ and } \frac{V}{V^*} = \frac{\rho^*}{\rho} = 0.091$$

Stagnation enthalpy rise due to the combusion of one kg of fuel is

$$\Delta h_o = \frac{41870}{29+1} = \frac{41870}{30} = 1395.66\ kJ/kg \text{ of air fuel mixture}$$

$$\Delta h_o = c_p (T_{o2} - T_{o1})$$

$$T_{o2} = 352.82 + 1388.72 = 1741.54\ K$$

$$\frac{T_{o2}}{T_o^*} = \frac{T_{o2}}{T_{o1}} \times \frac{T_{o1}}{T_o^*} = \frac{1741.54}{352.82} \times 0.174 = 0.8588$$

From Rayleigh tables

At $\frac{T_{o2}}{T_o^*} = 0.8588,\ M_2 = 0.64$ Ans

(b) From Rayleigh tables

At $M_2 = 0.64$

$$\frac{p_2}{p^*} = 1.525, \frac{p_{o2}}{p_o^*} = 1.061, \frac{T_2}{T^*} = 0.953,$$

and $$\frac{V}{V^*} = \frac{\rho^*}{\rho} = 0,625$$

$$p_2 = \frac{(p/p^*)_{M_2}}{(p/p^*)_{M_1}} \times p_1 = \frac{1.525}{2.273} \times 55 = 36.9\ k\,Pa \qquad \text{Ans}$$

$$T_2 = \frac{(T/T^*)_{M_2}}{(T/T^*)_{M_1}} \times T_1 = \frac{0.953}{0.207} \times 350 = 1611.35\ K \qquad \text{Ans}$$

$$V_2 = \frac{(V/V^*)_{M_2}}{(V/V^*)_{M_1}} \times V_1 = \frac{0.625}{0.091} \times 75 = 515.1\ m/s \qquad \text{Ans}$$

$$p_{o2} = \frac{(p_o/p_o^*)_{M_2}}{(p_o/p_o^*)_{M_1}} \times p_{o_1}$$

$$= \frac{1.061}{1.235} \times 56.526 = 48.562\ kPa \qquad \text{Ans}$$

(c) Stagnation pressure loss

$$\Delta p_o = p_{o1} - p_{o2} = 56.526 - 48.562 = 7.964\ kPa$$

Therefore

$$\frac{\Delta p_o}{p_{o1}} \times 100\% = \frac{7.965}{56.526} \times 100\% = 14.089\ \% \qquad \text{Ans}$$

(d) The maximum stagnation temperature for given inlet conditions exists at the sonic point, therefore

$$T_{omax} = T_{o2}^* = T_{o1}^* = T_o^*$$

Therefore, $\dfrac{T_{o1}}{T_o^*} = 0.174$

$$T_o^* = \frac{T_{o1}}{0.174} = \frac{352.82}{0.174} = 2027.7\ K$$

EXERCISES

A. THEORY

4.1 Write down the assumptions used for diabatic flow.

4.2 Define diabatic flow.

4.3 Draw Rayleigh curve and explain.

4.4 Draw the curve for diabatic flow parameters and explain the effect of properties on heating and cooling conditions.

4.5 Derive the Rayleigh flow relation.

4.6 What is diabatic flow ? Give two practical examples. Under what conditions the assumption of diabatic flow is not valid in a heat exchanger ?

4.7 Prove that the Mach number at the maximum enthalpy and maximum entropy points on the Rayleigh line are $1/\sqrt{k}$ and 1 respectively.

4.8 What is the value of Mach number of air at the maximum enthalpy point in Rayleigh heating process ? Depict static and stagnation values of pressure and temperature at this point and the sonic point on a Rayleigh line.

4.9 Draw a Rayleigh line on T-s plane. Show the following on this line. (a) stagnation temperature lines for both subsonic and supersonic flows, (b) maximum stagnation temperature point, (c) constant static and stagnation pressure lines.

4.10 Develop the following relations for diabatic flow.

(a) $$\frac{p_2}{p_1} = \frac{1 + k\,M_1^2}{1 + k\,M_2^2}$$

(b) $$\frac{p_{o_2}}{p_{o1}} = \frac{1 + k\,M_1^2}{1 + k\,M_2^2}\left[\frac{1 + \frac{k-1}{2}M_2^2}{1 + \frac{k-1}{2}M_1^2}\right]^{k/k-1}$$

(c) $$\frac{T_2}{T_1} = \left[\frac{M_2}{M_1}\left(\frac{1 + k\,M_1^2}{1 + k\,M_2^2}\right)\right]^2$$

(d) $$\frac{T_{o_2}}{T_{o1}} = \frac{T_2}{T_1}\left[\frac{1 + \frac{k-1}{2}M_1^2}{1 + \frac{k-1}{2}M_2^2}\right]$$

(e) $$\frac{T_{o_2}}{T_{o1}} = \left[\frac{M_2}{M_1}\left(\frac{1 + k\,M_1^2}{1 + k\,M_2^2}\right)\right]^2\left[\frac{1 + \frac{k-1}{2}M_1^2}{1 + \frac{k-1}{2}M_2^2}\right]$$

(f) $$\frac{T}{T^*} = \frac{M^2\,(1+k)^2}{(1 + k\,M^2)^2}$$

(g) $$\frac{T_o}{T_o^*} = \frac{2\,M^2\,(k+1)\left[1 + \frac{k-1}{2}M^2\right]}{(1 + k\,M^2)^2}$$

(h) $\frac{p}{p^*} = \frac{k+1}{1+k M^2}$

(i) $\frac{p_o}{p_o^*} = \frac{k+1}{1+k M^2}\left[\frac{2\left(1+\frac{k-1}{2}M^2\right)}{k+1}\right]^{k/k-1}$

(j) $\frac{s_2 - s_1}{R} = \ln \frac{M_2}{M_1}\left(\frac{M_2}{M_1} \times \frac{1+k M_1^2}{1+k M_2^2}\right)^{(k+1)/k-1}$

(k) $\frac{V_1}{V_2} = \frac{\rho_2}{\rho_1} = \left(\frac{M_1}{M_2}\right)^2 \frac{1+k M_2^2}{1+k M_1^2}$

4.11 Prove that the dimensionless value of the heat transfer in a Rayleigh process from a Mach number M_1 *to* M_2 is

$$\frac{Q}{c_p T_1} = \frac{M_2^2 - M_1^2}{2 M_1^2 (1+k M_2^2)^2} [2(1 - k M_1^2 M_2^2) + (k-1)(M_2^2 + M_1^2)]$$

Hence, show that the maximum value of heat transfer at a Mach numbr M is

$$\frac{Q_{max}}{c_p T_1} = \frac{(1-M)^2}{2(1+k) M^2}$$

B. UNSOLVED PROBLEMS

4.1 Air flows in a short tube without appreciable friction. Heat is supplied to increase the Mach number from an inlet value of 0.4 at a temperature of 35°C to be an exit value of 0.7. What is the heat flow per unit mass of air ? [229 kJ/kg]

4.2 Fuel is burned in a stream of air moving with a Mach number of 2, at a temperature of – 105°C and pressure of 0.68 bar in a duct of constant area. Estimate the maximum possible rise in stagnation temperature and corresponding values of pressure, temperature and stagnation pressure. [M = 1 for maximum possible stagnation temperature rise, $T_o^* = 380.8\ K$, $p_o^* = 3.52$, $T^* = 317.5\ k$ and $p^* = 1.86\ bar$].

4.3 Air flows without friction in a short section of constant area duct. At the duct inlet, $M_1 = 0.3$, $t_1 = 50°C$ and $\rho_1 = 2.16\ kg/m^3$. At the duct outlet, $M_2 = 0.6$. Determine the heat transfer per unit mass the entropy change and change in stagnation pressure for the process. [Δp_o = 22 kPa, 4.47 kJ/kg, 0.889 kJ/kg K].

4.4 Air flows without friction on constant area duct. At the entrance $M_1 = 2$, p_1 = 138 kPa and $t_1 = 38^oC$. As air moves along with the duct, heat is removed as long as it is possible to do so. Calculate the mass flow per unit area ($\dot{m}$ /A) and maximum velocity that is reached. [1093.8 kg/s m^2, 832 m/s]

4.5 Air flows without friction in a constant area duct. At section one, $M_1 = 0.5$, p_1 = 1.1 M Pa and T_{o1} = 335 K. Through the effect of heat transfer, the Mach number is raised to $M_2 = 0.9$ at section two. Determine the heat transfer per unit mass between sections one and two and the pressure p_2. [145 kJ/kg, 696 kPa].

4.6 A constant area duct is fed with air from a converging diverging nozzle. At the entrance to the duct the following properties are known p_{o1} = 800 *kPa*, T_{o1} = 700 *k* and M_1 = 3. A short distance down the duct (at section 2) the density is 0.334 kg/m^3. Assuming frictionless flow, determine the speed, pressure, and Mach number at section 2, and the heat transfer between the inlet and section 2. [866 m/s, 46.4 kPa, M_2 = 1.96, 156 kJ/kg].

4.7 Consider the frictionless flow of air in a constant area duct. At section 1, M_1 = 0.5, p_1 = 1.1 MPa, and T_{o1} = 333 K. Through the effect of heat transfer, the Mach number at section 2 is M_2 = 0.9 and the stagnation temperature. T_{o2} = 478 *K*.

(a) Determine the amount of heat transfer per unit mass to or from the fluid between section 1 and 2.

(b) Determine the pressure difference $p_1 - p_2$. [145 kJ/kg, 405 kPa].

4.8 At a position 3 m from the exit of a constant area duct A = 0.02 m^2, air is at an absolute pressure of 126 kPa and a temperature of 260°C. The air flows steadily through the duct (without friction) at the rate of 1.83 kg/s. The air leaves the duct subsonically at atmospheric pressure. Determine the Mach number, temperature, and stagnation temperature at the exit of the duct and the heat transfer over the 3 m of duct length. [T_2 = 1479 K, M_2 = 0.498, T_{o2} = 1552 K, 1.85 MJ/s].

4.9 Air flows without friction through a short duct of constant area. At the duct entrance, $M_1 = 0.3$, $t_1 = 50^oC$, and ρ_1 = 2.16 kg/m^3. As a result of heating the Mach number and density at the tube outlet are M_2 = 0.6 and ρ_2 = 0.721 kg/m^3. Determine the heat transfer per unit mass and the entropy change for the process. [449 kJ/kg, 0.892 kJ/kg K].

4.10 Air flows without friction through a duct of 80 mm diameter. At the duct inlet, the Mach number, stagnation temperature and absolute pressure are M_1 = 0.2, t_{o1} = 5°C, and p_1 = 2.75 bar. Heat is added at the rate of 220 kJ/kg of flowing fluid with the result that the temperature at the duct outlet is 490 K. Determine the stagnation temperature and Mach number at the duct outlet, and the entropy change for the process. [T_{o2} = 497 K, M_2 = 0.286, 0.593 kJ/kg K].

4.11 Air flows at the rate of 1.42 kg/s through a duct of 100 mm diameter. At the inlet section the temperature and absolute pressure are 52°C and 60 kPa, respectively. At the section downstream where the flow is choked, t_2 = 45°C. Determine the heat transfer unit mass, the entropy change and the change in stagnation pressure for the process assuming frictionless flow. [0.053 kJ/kg K, 18 kJ/kg, 2 kPa].

4.12 Using coordinates T/T^* and $(s - s^*)/c_p$ where s^* is the entropy at the condition where Mach number M = 1. Plot the Rayleigh line for air flow (k = 1.4) for Mach number in a range 0.4 < M < 3.

4.13 A gaseous mixture of air and fuel enters a ramjet combustion chamber with a velocity of 60 m/s and temperature of 50°C and at a pressure of 35 kPa. The heat of reaction of the mixture for the particular fuel to air ratio employed is 1160 kJ/kg. Find the conditions of stream of exit of combusion chamber if the friction is neglected and cross sectional area is constant. Also assume that the properties of both reactants and products of combusion are the same as air. [M_1 = 0.167, M_2 = 0.44, T_2 = 1447 K, p_2 = 0.286 bar, p_{o2} = 0.327 bar].

4.14 Air is brought into a tube through a converging-diverging nozzle designed for a Mach number of 2, and has initially stagnation temperature of 278 K and stagnation pressure of 680 kPa. The air is to be cooled by some mechanism not involving friction. Assume that there are no shocks, compute the maximum possible amount of heat rejection per kilogram of air passing through the duct, the final temperature, pressure and stagnation pressure. [105.6 kJ/kg, T_2 = 0, p_{o2} = ∞, p_2 = 0]

4.15 A gas at a pressure of 69 kPa and temperature 5°C enters a combustion chamber at a velocity of 60 m/s. The head supplied in the combusion chamber is 1406 kJ/kg. Determine the Mach number, pressure temperature and velocity of the gas at the exit. [M_2 = 0.64, p_2 = 45.8 kPa, T_2 = 1549.3 K, V_2 = 506 m/s]

4.16 The stagnation temperature of air in a combustion chamber is increased to 3.5 times its initial value. If the air at entry is at 500 kPa, 378 K and a Mach number of 0.25 determine (a) the Mach number, pressure and temperature at the exit; (b) stagnation pressure loss, and (c) the heat supplied per kg of air. [M_2 = 0.68, p_2 = 330 kPa, 944°C Δp_o = 73.5 kPa, 962.3 kJ/kg]

4.17 Air at 300 kPa, 15°C and M_1 = 1.5 is brought to sonic velocity in a frictionless constant area duct through which heat transfer can occur. (a) Determine the final pressure and temperature and the heat added during the process. (b) What will be the Mach number, pressure and temperature of air if this heat is extracted from the air ? [519 kPa, 382.47 K, 42.05 kJ/kg, M_2 = 1.88, 209 kPa, 219.92 K].

5

FANNO FLOW

5.1 INTRODUCTION

The flow of compressible gases in ducts of constant area is of importance in many engineering fields, including stationary power plants, transport of fluids in chemical plants, aircraft propulsion engines, high vacuum technology, various types of flow machinery and transport of natural gas in long pipe lines.

We shall here consider flows in which wall friction is the chief factor bringing about changes in fluid properties.

5.2. Fanno flow

In the last chapter, certain assumptions were made which led to a type of flow which was called Rayleigh flow. In this article, too, certain assumptions will be made which, when followed, lead to a pattern known as Fanno flow. The requirements for Fanno flow are that :

(1) Flow must be steady

(2) Flow is one dimensional

(3) Flow is adiabatic (no heat transfer)

(4) Flow takes place in a constant area section.

(5) The gas is perfect and hence has constant specific heats

(6) No work transfer,

5.2.1 The Fanno curves

Frictional, steady flow which takes place in constant area duct in the absence of heat transport termed Fanno flow. Further more, equations which are applicable to such flow are called, Fanno equations.

The equation of continuity is written as (same as equation 4.1)

$$\frac{\dot{m}}{A} = \rho V = G$$

and the energy equation

$$h + \frac{V^2}{2} = h_o \tag{5.1}$$

Combining equations (4.1) and (5.1), we get the equation of Fanno line

$$h = h_o - \frac{G^2}{2\rho^2} \tag{5.2}$$

Since h_o and G are constant for given flow, equation (5.2) defines a relation between the local enthalpy and local density.

For a pure substance, the entropy, can be expressed, as a function of two properties e.g.

$$s = f(h, \rho)$$

or, $$\rho = f(s, h) \tag{5.3}$$

Combining of equations (5.2) and (5.3) yields also the equation of the Fanno line

$$h = h_o - \frac{G^2}{2\,[f(s, h)]^2} \tag{5.4}$$

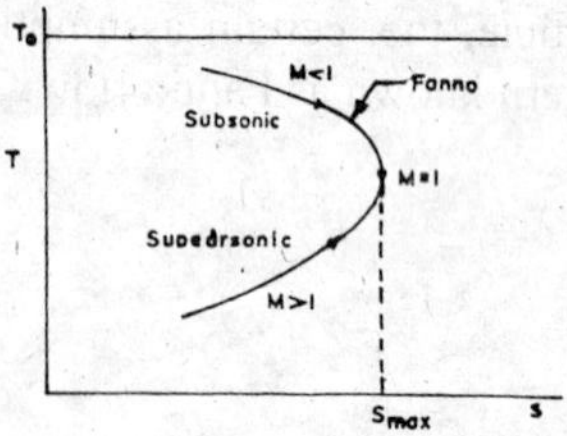

Fig 5.1 Fanno line

With the help of equation (5.2) or (5.4) for given values of h_o and G a curve in the h–s plane as shown in Fig (5.1) is obtained.

The upper part of Fanno curve corresponds to subsonic flow, and the lower branch corresponds to supersonic flow, and that the Mach number is unity at the point of maximum entropy on Fanno curve.

From the Gibbs's relation (From Themodynamics)

$$T\,ds = dh - \frac{dp}{\rho} \tag{5.5}$$

Differentiating equation (5.2)

$$dh = G^2 \frac{d\rho}{\rho^3}$$

so that

$$T\,ds = G^2 \frac{d\rho}{\rho^3} - \frac{dp}{\rho} \tag{5.6}$$

At the point of maximum entropy on a Fanno curve ds = 0 and the above equation (5.6) gives

$$G^2 \frac{d\rho}{\rho^3} = \frac{dp}{\rho}$$

or, $$\frac{\rho^2 V^2}{\rho^3} d\rho = \frac{dp}{\rho}$$

or, $$V^2 = \frac{dp}{d\rho}$$

or, $$\frac{V^2}{a^2} = 1$$

or, $$M^2 = 1$$

or, $$M = 1$$

Therefore M = 1 at the point of maximum entropy.

Since the stagnation temperature is the same for all points on the Fanno line, the temperature at point a (Fig 5.1) is the critical temperature, T^* for the entire Fanno line. Thus, fanno flow corresponding to the portion of the Fanno line above the critical temperature must be subsonic, and Fanno flow on the line below T^* must be supersonic (Fig 5.1)

For Fanno flow to be consistent with the second law of thermodynamics, flow can proceed along the Fanno line toward state a, the critical state. The critical state may or may not be reached by the flow. If it is, the Fanno flow is choked. Fanno flow behaviour are summarized in Fig 5.2. A case involving subsonic Fanno flow that is accelerated by friction to a higher Mach number without choking is shown in fig 5.2(a). A supersonic flow that is decelerated by friction to a lower Mach number without choking is shown in Fig 5.2(b). In Fig 5.2(c), an abrupt change from supersonic to subsonic flow in the Fanno duct is represented. This sudden deceleration occurs across a standing normal Shock wave. For details see the next chapter.

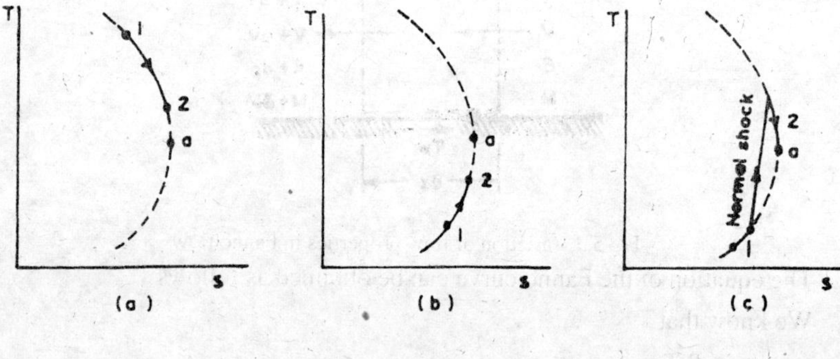

Fig 5.2. Fanno flow behaviour

For Fig. 5.2 (a)

(b)

(c)

5.2.2 Important features of Fanno curve

(1) From the second law of thermodynamics, the entropy of the adiabatic flow increases but not decreases. Thus, the path of states along the Fanno curve must be toward the right.

(2) In the subsonic region, the effects of friction will be to increase the velocity and Mach number and to decrease the enthalpy and pressure of the stream.

(3) In the supersonic region, the effects of friction will be to decrease the velocity and Mach number and to increase the enthalpy and pressure of the stream.

(4) A subsonic flow can never become supersonic, due to the limitation of second law of thermodynamics, but in can approach to sonic i.e. M = 1.

(5) A supersonic flow can never become subsonic, unless a discontinuity (shock) is present.

(6) In the case of isentropic stagnation, pressure is reduced whether the flow is subsonic or supersonic.

5.2.3 Fanno flow relations for a perfect gas

Fig 5.3 shows the variation of flow properties along a duct in Fanno flow processes. The variation of flow properties are due to friction.

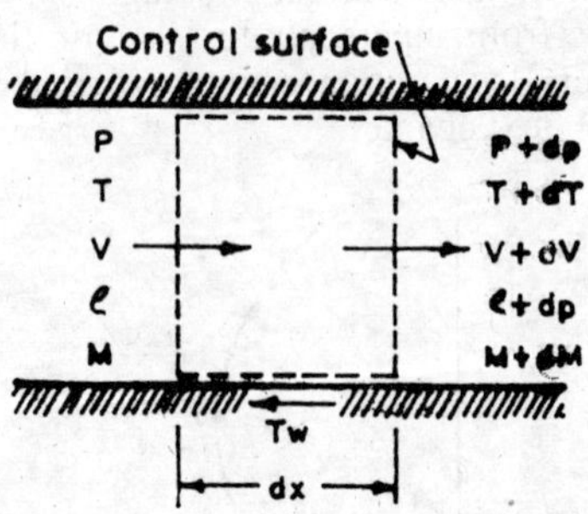

Fig 5.3 Variation of flow properties in Fanno flow.

The equation of the Fanno curve can be obtained as follows

We know that

$$p = \rho R T$$

Taking log on both the sides, we get

$$\frac{dp}{p} = \frac{d\rho}{\rho} + \frac{dT}{T} \tag{5.7}$$

we also know that

$$M = \frac{V}{a} = \frac{V}{\sqrt{k\,R\,T}}$$

$$M^2 = \frac{V^2}{k\,R\,T}$$

Taking log on both the sides, we get

$$\frac{dM^2}{M^2} = \frac{dV^2}{V^2} - \frac{dT}{T}$$

The energy equation of steady flow is written for a perfect gas as

$$h + \frac{V^2}{2} = constant$$

Differentiating by parts

$$dh + VdV = 0$$

or, $$c_p\, dT + VdV = 0$$

Dividing by $c_p\, dT$, we get

$$1 + \frac{VdV}{c_p dT} = 0 \tag{5.8}$$

Combining equation (5.7) and (5.8) and using the relation $c_p = \frac{kR}{k-1}$, we get

$$\frac{dT}{T} + \frac{(K-1)}{K\,R\,T}\frac{dV^2}{2} = 0$$

or, $$\frac{dT}{T} + \left(\frac{k-1}{2}\right) M^2 \frac{dV^2}{V^2} = 0 \tag{5.9}$$

The continuity equation is written as

$$\frac{\dot{m}}{A} = G = \rho\, V$$

Taking log on both sides

$$\ln G = \ln \rho + \ln V$$

Differentiating

$$\frac{d\rho}{\rho} + \frac{dV}{V} = 0$$

or, $$\frac{d\rho}{\rho}+\frac{1}{2}\left(\frac{dV^2}{V^2}\right)=0 \tag{5.10}$$

Substituting the value of $\left(\frac{dT}{T}\right)$ and $\left(\frac{d\rho}{\rho}\right)$ from the equation (5.9) and (5.10) into equation (5.7)

$$\frac{dp}{p}=-\frac{1}{2}\left(\frac{dV^2}{V^2}\right)-\left(\frac{k-1}{2}\right)M^2\frac{dV^2}{V^2}$$

$$=-\frac{dV^2}{V^2}\left[\frac{1}{2}+\frac{k-1}{2}M^2\right]$$

$$=-\frac{dV^2}{V^2}\left[\frac{1+(k-1)M^2}{2}\right]$$

$$\frac{dV^2}{V^2}=-\frac{dp}{p}\left[\frac{2}{1+(k-1)M^2}\right] \tag{5.11}$$

From Fig (5.3), the momentum equation is written as

$$pA-(p+dp)A-\tau_w\,dA_w=\rho AV(V+dV-V)$$

or, $$-A\,dp-\tau_w\,dA_w=\rho AV\,dV \tag{5.12}$$

Where A is the cross sectional area, dA_w is wetted wall area and τ_w wall shear stress.

We know that the wall shear stress τ_w and friction factor λ is related as

$$\tau_w=\lambda\frac{\rho V^2}{8}$$ [We are using $\lambda = 4f$]

The wall shear stress is written as

$$dA_w=\text{perimeter}\times\text{length}=\pi D\,dx \tag{5.13}$$

$$-A\,dp-\frac{\lambda\rho V^2}{8}(\pi D\,dx)=\rho AV\,dV$$

Dividing the above equation by A, we get

$$-dp-\frac{\lambda\rho V^2\pi D\,dx}{A}=\rho V\,dV$$

or, $$dp+\frac{\rho V^2}{2}\left(\frac{\lambda\,dx}{D}\right)+\rho V\,dV=0$$

or, $$dp+\frac{\rho V^2}{2}\left(\frac{\lambda\,dx}{D}\right)+\frac{\rho V^2}{V}dV=0$$

Next, we divide by p, and upon noting that $\rho V^2 = k\,p\,M^2$, we get

$$\frac{dp}{p} + \frac{k\,M^2}{2}\left(\frac{\lambda\,dx}{D}\right) + \frac{k\,M^2}{2}\left(\frac{d\,V^2}{V^2}\right) = 0 \qquad (5.14)$$

The relation between stagnation and static pressure is given by

$$\frac{p_o}{p} = \left(1 + \frac{k-1}{2}M^2\right)^{k/(k-1)}$$

Taking log on both the sides

$$\ln p_o = \ln p + \frac{k}{k-1}\ln\left(1 + \frac{k-1}{2}M^2\right)$$

Differentiating on both the sides

$$\frac{dp_o}{p_o} = \frac{dp}{p} + \frac{k}{k-1}\left[\frac{d\left(1 + \frac{k-1}{2}M^2\right)}{1 + \frac{k-1}{2}M^2}\right]$$

or,

$$\frac{dp_o}{p_o} = \frac{dp}{p} + \frac{k\,M^2}{2\left(1 + \frac{k-1}{2}M^2\right)}\frac{d\,M^2}{M^2} \qquad (5.15)$$

Definition of impulse function gives

$$F = p\,A + \rho\,A\,V^2 = p\,A + k\,P\,A\,M^2$$

By logarthmic differentiation

$$\frac{dF}{F} = \frac{d\,(1 + k\,M^2)}{1 + k\,M^2} + \frac{dp}{p}$$

or,

$$\frac{dF}{F} = \frac{k\,M^2}{1 + k\,M^2}\left(\frac{dM^2}{dM^2}\right) + \frac{dp}{p} \qquad (5.16)$$

.2.4 Effects of wall friction on fluid properties

At this point we notice that we have seven simultaneous, linear algebraic quations, namely, equations (5.7), (5.8), (5.10), (5.11) (5.14) (5.15) and (5.16), which relate eight differential variables $\frac{dp}{p}, \frac{d\rho}{\rho}, \frac{dT}{T}, dM^2/M^2, dV^2/V^2, dp_o/p_o, \frac{dF}{F}$ and $\lambda\,dx/D$. The physical henomenon causing changes in state in viscous friction. Hence we select the ariable λ dx/D as independent. The remaining seven variables may onsequently be found in terms of λ dx/D with the aid of the seven equations. The usual method for solving simultaneous equations are used. For example, the erm dT/T may be eliminating from equation (5.7) and (5.10), yielding

$$\frac{dp}{p} = \frac{d\rho}{\rho} - \frac{k-1}{2} M^2 \frac{dV^2}{V^2}$$

Next, $\frac{d\rho}{\rho}$ is eliminated from this expression with the help of equation (5.10), thus giving

$$\frac{dp}{p} = -\frac{1+(k-1)M^2}{2}\left(\frac{dV^2}{V^2}\right)$$

Finally, we use this relation for eliminating dV^2/V^2 from equation (5.14) and obtain, following algebraic rearrangement.

$$\frac{dp}{p} = \frac{-kM^2\left(1+\frac{(k-1)}{2}M^2\right)}{2(1-M^2)}\left(\lambda\frac{dx}{D}\right) \tag{5.17}$$

Using similar methods, the following formulas are obtained.

$$\frac{dM^2}{M^2} = \frac{kM^2\left(1+\frac{k-1}{2}M^2\right)}{(1-M^2)}\left(\lambda\frac{dx}{D}\right) \tag{5.18}$$

$$\frac{dV}{V} = \frac{kM^2}{2(1-M^2)}\left(\lambda\frac{dx}{D}\right) \tag{5.19}$$

$$\frac{dT}{T} = \frac{1}{2}\left(\frac{dV}{V}\right) = -\frac{k(k-1)M^4}{2(1-M^2)}\left(\lambda\frac{dx}{D}\right) \tag{5.20}$$

$$\frac{d\rho}{\rho} = -\frac{kM^2}{2(1-M^2)}\left(\lambda\frac{dx}{D}\right) \tag{5.21}$$

$$\frac{dp_o}{p_o} = -\frac{kM^2}{2}\left(\lambda\frac{dx}{D}\right) \tag{5.22}$$

$$\frac{dF}{F} = -\frac{kM^2}{2(1+kM^2)}\left(\lambda\frac{dx}{D}\right) \tag{5.23}$$

It appears from equations (5.22) and (5.23) that both the isentropic stagnation pressure and impulse function must decrease when friction is present. This is true for either subsonic or supersonic flow. Wall friction therefore reduces the effectiveness of all types of flow machinery and also reduces the thrust obtainable from jet propulsion devices.

The direction of changes in the remaining stream properties, according to equations (5.17) through (5.21), depend on whether the Mach number is greater or less than unity, since the term $(1 - M^2)$ appears in the denominator of each of these equations. We may summarize the various changes in table 5.1.

TABLE 5.1 Summary of Fanno flow behaviour

	Parameter	Flow M < 1	Flow M > 1
1	Pressure, p	Decrease	Increase
2	Mach number, M	Increase	Decrease
3	Velocity, V	Increase	Decrease
4	Temperature, T	Decrease	Increase
5	Density, ρ	Decrease	Increase
6	Stagnation pressure, p_o	Decrease	Decrease
7	Stagnation temperature T_o	Constant	Constant
8	Impulse Function	Decrease	Decrease
9	Friction	Accelerates flow	Decelerates flow

We see that the Mach number always tends towards unity. Continuous transitions either from subsonic to supersonic flow, or from supersonic to subsonic, are consequently impossible.

5.2.5 Working formula for Fanno flow

In this article we will integrate the differential equations to obtain formulas suitable for practical compulations. We will use the Mach number as the independent variable for this purpose.

Equation (5.18) is rearranged to need

$$\int_o^{L_{max}} \frac{\lambda\, dx}{D} = \int_{M^2}^{l} \frac{(1-M^2)}{k\,M^4\left(1+\dfrac{k-1}{2}M^2\right)}\cdot dM^2$$

where the limits of integration are taken as (i) the section where the Mach number is M, and where x is arbitrarily set equal to zero, and (ii) the section where the Mach number is unity, and x is the maximum possible length of duct, L_{max}.

Therefore,

$$\lambda_{av}\frac{L_{max}}{D} = \frac{1-M^2}{k\,M^2} + \frac{k+1}{2k}\ln\frac{(k+1)\,M^2}{2\left(1+\dfrac{k-1}{2}M^2\right)} \qquad (5.24)$$

Where λ_{av} is the mean friction coefficient with respect to length, defined by

$$\lambda_{av} = \frac{1}{L_{max}}\int_o^{L_{max}} \lambda\, dx$$

Equation (5.24) gives the maximum value of λ_{av} L/D corresponding to any initial Mach number M.

Since $\lambda_{av} L_{max}/D$ is a function of only of M, the length of duct L required for the flow to pass from a given initial Mach Number M_1 to a given final Mach number M_2 is found from the expression

$$\lambda_{av}\frac{L}{D} = \left(\lambda_{av}\frac{L_{max}}{D}\right)_{M_1} - \left(\lambda_{av}\frac{L_{max}}{D}\right)_{M_2} \tag{5.25}$$

Local stream properties are found in the local Mach number.

Equation (5.17) and (5.18) may be combine to give

$$\frac{dp/p}{dM^2/M^2} = -\frac{1+(k-1)M^2}{2\left(1+\frac{k-1}{2}M^2\right)}$$

or,
$$\frac{dp}{p} = -\frac{1+(k-1)M^2}{2M^2\left(1+\frac{k-1}{2}M^2\right)}dM^2$$

Denoting the pressure at M = 1 by the symbol p^*, we integrate between the section where M = M, p = p and the section where M = 1, $p = p^*$, and thus obtain

$$\frac{p}{p^*} = \frac{1}{M}\sqrt{\frac{k+1}{2\left(1+\frac{k-1}{2}M^2\right)}} \tag{5.26}$$

By similar methods we get the remaining relations.

$$\frac{V}{V^*} = M\sqrt{\frac{k+1}{2\left(1+\frac{k-1}{2}M^2\right)}} \tag{5.27}$$

$$\frac{T}{T^*} = \frac{a^2}{a^{*2}} = \frac{k+1}{2\left(1+\frac{k-1}{2}M^2\right)} \tag{5.28}$$

$$\frac{\rho}{\rho^*} - \frac{V^*}{V} = \frac{1}{M}\sqrt{\frac{2\left(1+\frac{k-1}{2}M^2\right)}{k+1}} \tag{5.29}$$

$$\frac{p_o}{p_o^*} = \frac{1}{M}\sqrt{\left[\frac{2\left(1+\frac{k-1}{2}M^2\right)}{(k+1)}\right]^{(k+1)/(k-1)}} \tag{5.30}$$

$$\frac{F}{F^*} = \frac{1 + kM^2}{M\sqrt{2(k+1)\left(1 + \frac{k-1}{2}M^2\right)}} \tag{5.31}$$

In order to find the same in some stream property, say pressure, between the sections where the Mach numbers are M_1 and M_2, respectively, we get

$$\frac{p_2}{p_1} = \frac{(p/p^*)_{M_2}}{(p/p^*)_{M_1}}$$

$$= \frac{M_1}{M_2}\sqrt{\frac{1 + \frac{k-1}{2}M_1^2}{1 + \frac{k-1}{2}M_1^2}} \tag{5.32}$$

Similarly we obtain the remaining variables.

5.2.6 Working tables and charts

It is convenient to calculate the above dimensionless ratio for various Mach numbers for use in solving numerical problems. A convenient reference state for this purpose is where M = 1; an asterisk is used to denote this state. It is to be noted that this is a different state from that used in the isentropic flow tables where the condition M = 1 is reached with the entropy kept constant.

The various property ratios as obtained from the Fanno line table have been plotted in Fig 5.4.

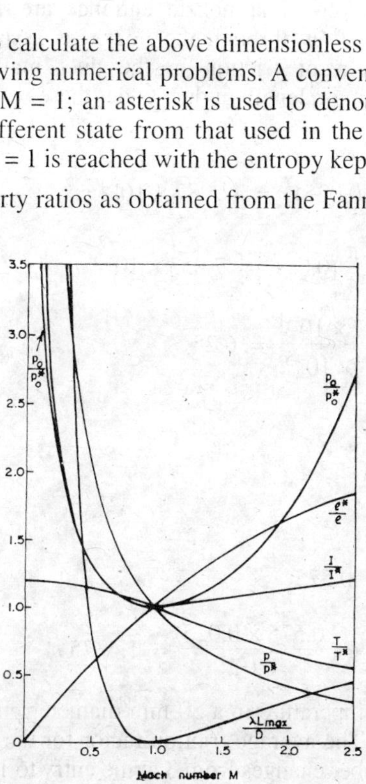

Fig 5.4 Function for constant area flow with friction (k = 1.4)

ILLUSTRATIVE EXAMPLES

5.1 The Mach number for air flow in a pipe increases from 0.3 to 0.7 because of pipe friction. The initial pressure is 140 kPa. What is the pressure at the section where Mach number is 0.7 if the flow is adiabatic ?

Solution From Fanno tables

At $M_1 = 0.3, \quad p_1/p^* = 3.619$

At $M_2 = 0.7; \quad p_2/p^* = 1.493$

$$\text{Therefore, } p_2/p_1 = \frac{(p/p^*)\, M_2}{(p/p^*)\, M_1} = \frac{1.493}{3.619} = 0.4125$$

$$p_2 = 0.4125\, p_1 = 0.4125 \times 140 = 57.75 \text{ kPa}$$ Ans

5.2 Air enters an insulated tube 25 mm diameter through a convergent divergent nozzle with a throat diameter of 10 mm. The pressure and temperature in the low velocity region at nozzle entrance are 700 kPa and 333 K respectively. The flow through the nozzle is assumed frictionless and average coefficient of friction in the duct may be taken as 0.01. Determines maximum length of duct under these conditions.

Solution

$$A = \frac{\pi}{4} D^2 = \frac{\pi}{4} (0.025)^2 = 4.9087 \times 10^{-4}\, m^2$$

$$A^* = \frac{\pi}{4} D^{*2} = \frac{\pi}{4} (0.01)^2 = 7.853 \times 10^{-5}\, m^2$$

$$\text{Therefore, } \frac{A}{A^*} = \frac{4.9087 \times 10^{-4}}{7.853 \times 10^{-5}} = 6.25$$

From isentropic tables,

At $A/A^* = 6.24,\ M \approx 3.4$

From Fanno tables

$$\text{At M} = 3.4, \frac{\lambda_{av} L_{max}}{D} = 0.575$$

$$L_{max} = \frac{0.575\, D}{\lambda_{av}} = \frac{0.575 \times 0.025}{0.01} = 1.4375\, m$$ Ans

5.3 The pressure and temperature in a 50 mm diameter pipe are 10 bar and 400 K, respectively. The average friction factor for the pipe surface is 0.008. If the Mach number changes from 3 at the entry to 1 at the exit determine.

(a) the length of the pipe and (b) the mass flow rate.

Solution

(a) From Fanno tables

At $M_1 = 3, \dfrac{\lambda_{av} L_{max}}{D} = 0.522$

and at $M_1 = 1, \dfrac{\lambda_{av} L_{max}}{D} = 0$

$$\text{Therefore, } \frac{\lambda_{av} L}{D} = \left(\frac{\lambda_{av} L_{max}}{D}\right)_{M_1} - \left(\frac{\lambda_{av} L_{max}}{D}\right)_{M_2}$$

$$= 0.522 - 0 = 0.522$$

$$\text{Therefore } L = \frac{0.522 \times 0.05}{0.008} = 3.2625\ m$$ Ans

(b) From isentropic flow tables

At $M_1 = 3, \dfrac{T_1}{T_o} = 0.357, \dfrac{p_1}{p_o} = 0.0272, \dfrac{A_1}{A^*} = 4.235$

$$A_1 = \frac{\pi}{4}(0.05)^2 = 1.9635 \times 10^{-3}\ m^2$$

$$A^* = \frac{A_1}{4.235} = \frac{1.9635 \times 10^{-3}}{4.235} = 4.636 \times 10^{-4}\ m^2$$

$$p_1 = 0.0272 \times 10 = 0.272\ bar$$

$$T_1 = 0.357 \times 400 = 142.8\text{K}$$

$$p_1 = \rho_1 RT_1$$

$$0.272 \times 100 = \rho_1 \times 0.287 \times 142.8$$

$$\rho_1 = 0.66368\ kg/m^3$$

$$a_1 = \sqrt{k\,RT_1} = \sqrt{1.4 \times 287 \times 142.8} = 239.535\ m/s$$

$$V_1 = M_1 a_1 = 3 \times 239.535 = 718.6\ m/s$$

$$\dot{m} = \rho_1 A V_1 = 0.66368 \times 1.9635 \times 10^{-3} \times 718.6$$

$$= 0.93644\ \text{kg/s}$$

From the choked flow.

$$\frac{\dot{m}\sqrt{T_o}}{A^* p_o} = 0.040418$$

Therefore, $\dot{m} = \dfrac{0.040418 \times 4.636 \times 10^{-4} \times 10 \times 10^5}{\sqrt{400}}$

$= 0.9368$ kg/s

Ans

5.4 An experimental ring comprising a converging-diverging nozzle attached to a smooth round tube, the following data were obtained. The upstream pressure and temperature are 680 kPa and 315 K respectively. The throat diameter is 6.15 mm. The diameter of the circular tube and nozzle exit is 12.7 mm. The pressure of stream at a point which is 1.75 D of the tube is 24 kPa and at a point 30 D of the tube is 48 kPa.

Solution

It is desired to calculate average friction coefficients between section $x/D = 1.75$ and $x/D = 30$, by assuming that the flow to the throat of the nozzle is isentropic and the flow in remaing system is adiabatic.

$$\left(\frac{A}{A^*}\right)_1 \left(\frac{p}{p_o}\right)_1 = \left(\frac{12.7}{6.15}\right)^2 \times \frac{24}{680} = 0.15$$

At this point Mach no, is $M_1 = 2.54$

Also, $$\left(\frac{A}{A^*}\right)_2 \left(\frac{p}{p_o}\right)_2 = \left(\frac{12.7}{6.15}\right)^2 \times \frac{48}{680} = 0.3$$

At this point Mach no. is $M_2 = 1.58$

From Fanno tables

At $\quad M_1 = 2.54, \quad \dfrac{\lambda_{av} L_{max}}{D} = 0.440$

At $\quad M_2 = 1.58, \quad \dfrac{\lambda_{av} L_{max}}{D} = 0.165$

Therefore, $$\frac{\lambda_{av} L}{D} = \left(\frac{\lambda_{av} L_{max}}{D}\right)_{M_1} - \left(\frac{\lambda_{av} L_{max}}{D}\right)_{M_2}$$

$= 0.44 - 0.165 = 0.275$

Given $\quad \dfrac{L}{D} = \dfrac{(30 - 1.75)\, D}{D} = 28.25$

Therefore, $\lambda = \dfrac{0.275}{28.25} = 9.73 \times 10^{-3}$

Ans

5.5 The pressure and temperature at the inlet of a long straight smooth duct are 101 kPa and 298 K respectively. The flow is considered to be adiabatic and the pipe is long enough that the flow is choked. Find the speed and temperature at the pipe exit. If the duct diameter and pressure at inlet are 12 mm and 94.1 kPa respectively, find the duct length. Take λ_{av} = 0.0185.

Solution

(a) Since T_o = constant and $\frac{T_o}{T} = 1 + \frac{k-1}{2} M^2$

Therefore, $T_e = \frac{T_o}{1 + \frac{k-1}{2} M_e^2} = \frac{298}{1 + 0.2\,(1)^2} = 248.34\ K$ Ans

$$a_e = \sqrt{k\,RT_e} = \sqrt{1.4 \times 287 \times 248.34} = 315.88\ m/s$$

$V_e = M_e\,a_e = 1 \times 315.88 = 315.88\ m/s$ Ans

(b) $\frac{p_1}{p_o} = \frac{94.1}{101} = 0.93168$

From isentropic tables

At $\frac{p_1}{p_o} = 0.93168 \approx 0.932, \quad M_1 = 0.32 \quad \frac{T_1}{T_o} = 0.979,$

Therefore, $T_1 = 298 \times 0.979 = 291.742\ K$

From Fanno tables

At $M_1 = 0.32, \quad \frac{\lambda_{av}\,L_{max}}{D} = 4.447$

and $M_2 = 1, \quad \frac{\lambda_{av}\,L_{max}}{D} = 0$

Therefore, $\frac{\lambda_{av}\,L}{D} = \left(\frac{\lambda_{av}\,L_{max}}{D}\right)_{M_1} - \left(\frac{\lambda_{av}\,L_{max}}{D}\right)_{M_2}$

or, $\frac{\lambda_{av}\,L}{D} = 4.447 - 0 = 4.447$

$L = \frac{4.447 \times 12 \times 10^{-3}}{0.0185} = 2.88454\ m$ Ans

5.6 Air is drawn from the atmosphere at 20°C and 101 kPa through a converging nozzle into a long insulated 20 mm diameter tube of constant area. Flow in the nozzle is isentropic. The Mach number at the inlet to the constant area tube is 0.15. Evaluate the mass flow rate through the tube. Calculate T^* and p^* for the isentropic process. Calculate T^* and p^* for flow through the constant area tube. Show how these results would be changed if the Mach number at the tube inlet were raised to 0.3

Solution

Since flow in nozzle is isentropic, therefore,

$$T_{o1} = T_o \text{ and } p_{o1} = p_o$$

$$\text{At} \quad M_1 = 0.15 \; \frac{p_1}{p_{o1}} = 0.9744, \text{ and } \frac{T_1}{T_{o1}} = 0.995$$

$$\text{and at } M_2 = 1, \frac{T_2}{T_{o2}} = 0.8333, \text{ and } \frac{p_2}{p_o} = 0.5283$$

$$T_1 = 0.995 \times 293 = 291.535\ K$$

$$\text{Therefore, } p_1 = 0.9744 \times 101 = 98.41\ kPa$$

$$T_2 = 0.8333 \times 293 = 244.1569\ K = T^*$$

$$p_2 = 0.5283 \times 101 = 53.35831\ kPa = p^* \qquad \text{Ans}$$

From Fanno tables

$$\text{At} \quad M_1 = 0.15, p_1/p^* = 7.319$$

$$\text{Therefore, } p^* = \frac{p_1}{7.319} = \frac{98.41}{7.319} = 13.445\ kPa \qquad \text{Ans}$$

$$\text{Now, } p_1 = \rho_1 RT_1$$

$$98.41 = \rho_1 \times 0.287 \times 291.535$$

$$\rho_1 = 1.176\ kg/m^3$$

$$a_1 = \sqrt{k\ RT_1} = \sqrt{1.4 \times 287 \times 291.535} = 342.25\ m/s$$

$$V_1 = M_1 a_1 = 0.15 \times 342.25 = 51.338\ m/s$$

$$\text{Therefore } \dot{m} = \rho_1 A_1 V_1$$

$$= 1.176 \times \frac{\pi}{4}(0.02)^2 \times 51.338 = 0.01896\ kg/s \qquad \text{Ans}$$

If $M_1 = 0.3$, flow in tube starts on a different Fanno line.

5.7 Air flows steadily and adiabatically from a large tank through a converging nozzle connected to a constant area duct. The nozzle may be considered frictionless. Air in the tank is at p = 10 bar, t = 125°C. The absolute pressure at the nozzle exit (duct inlet) is 784 kPa. Determine the pressure at the end of the duct length, L, if the temperature there is 65°C. Also find the length, L, of 50 mm diameter between section one and two. Take $\lambda_{av} = 0.0192$.

Solution

$$\frac{p_1}{p_o} = \frac{784}{1000} = 0.784$$

From isentropic tables

At $\frac{p_1}{p_o} = 0.784, M_1 = 0.6, \frac{T_1}{T_o} = 0.933$

Therefore, $T_1 = 398 \times 0.933 = 371.334\ K$

From Fanno tables

At $M_1 = 0.6, \frac{T_1}{T^*} = 1.119\ \frac{p_1}{p^*} = 1.763$

Therefore, $T^* = \frac{371.334}{1.119} = 331.84\ K$

$$p^* = \frac{784}{1.763} = 444.7\ kPa$$

Now, $T_2/T^* = \frac{338}{331.84} = 1.01856$

From Fanno tables

At $T_2/T^* = 1.01856,\ M_2 \approx 0.94, \frac{p_2}{p^*} = 1.074$

Therefore, $p_2 = 1.076 \times p^* = 1.076 \times 444.7 = 478.5\ kPa$ Ans

From Fanno tables

At $M_1 = 0.6, \left(\frac{\lambda_{av} L_{max}}{D}\right)_{M_1} = 0.491$

and $M_2 = 0.94, \left(\frac{\lambda_{av} L_{max}}{D}\right)_{M_2} = 0.005$

Therefore, $\frac{\lambda_{av} L}{D} = \left(\frac{\lambda_{av} L_{max}}{D}\right)_{M_1} - \left(\frac{\lambda_{av} L_{max}}{D}\right)_{M_2}$

$$= 0.491 - 0.005 = 0.486$$

$L = \frac{0.486 \times D}{\lambda_{av}} = \frac{0.486 \times 50}{1000 \times 0.0192} = 1.265\ m$ Ans

5.8 Air flows in an insulated duct with a speed of 140 m/s. The temperature and pressure are 473 K and 20 bar respectively.

(a) Find the temperature in this duct where the pressure has dropped to 1.26 MPa as a result of friction.

(b) If the duct has a diameter of 150 mm, find the distance between the two points. Take friction factor $\lambda_{av} = 0.0147$.

Solution

(a) $$a_1 = \sqrt{k\,RT_1} = \sqrt{1.4 \times 287 \times 473} = 435.95\ m/s$$

$$M_1 = \frac{V_1}{a_1} = \frac{140}{435.95} = 0.32$$

From Fanno tables,

At $M_1 = 0.32,\ p_1/p^* = 3.388$

$$T_1/T^* = 1.176, \left(\frac{\lambda_{av} L_{max}}{D}\right)_{M_1} = 4.447$$

$$T^* = \frac{T_1}{1.176} = \frac{473}{1.176} = 402.2\ K$$

$$p^* = \frac{p_1}{3.388} = \frac{2}{3.388} = 0.59\ M\,Pa$$

Therefore, $\dfrac{p_2}{p^*} = \dfrac{1.26}{0.59} = 2.134$

From Fanno tables

At $p_2/p* = 2.134,\ M_2 = 0.5$

$$\frac{T_2}{T^*} = 1.143, \left(\frac{\lambda_{av} L_{max}}{D}\right)_{M_2} = 1.069$$

Therefore, $T_2 = 1.143 \times T^* = 1.143 \times 402.2 = 459.71\ K$ Ans

$$\frac{\lambda_{av} L}{D} = \left(\frac{\lambda_{av} L_{max}}{D}\right)_{M_1} - \left(\frac{\lambda_{av} L_{max}}{D}\right)_{M_2}$$

$$= 4.447 - 1.069 = 3.378$$

$$L_{12} = \frac{3.378 \times D}{\lambda_{av}} = \frac{3.378 \times 0.15}{0.0147} = 34.47\ m$$ Ans

5.9 A circular duct passes 8.25 kg/s of air at an exit Mach number of 0.5. The entry pressure and temperature are 345 kPa and 311 K respectively. The average friction factor is 0.02. If the Mach number at entry is 0.15, determine,

(a) the diameter of the duct,

(b) length of the duct

(c) pressure and temperature at the exit of duct, and

(d) stagnation pressure loss.

Verify the exit Mach number through exit velocity and temperature.

Solution

(a) $p_1 = \rho_1 RT_1$

$345 = \rho_1 \times 0.287 \times 311$

$\rho_1 = 3.865\ kg/m^3$

$a_1 = \sqrt{kRT_1} = \sqrt{1.4 \times 287 \times 311} = 353.4965\ m/s$

$V_1 = M_1 a_1 = 0.15 \times 353.4965 = 53.024\ m/s$

$\dot{m} = \rho_1 A_1 V_1$

$8.25 = 3.865 \times A_1 \times 53.024$

$A_1 = 0.04025\ m^2$

$\frac{\pi}{4} D^2 = 0.04025$

$D^2 = 0.05125$

$D = 0.226$ m $= 226$ mm Ans

From isentropic tables

At $M_1 = 0.15, \frac{p_1}{p_{o1}} = 0.984$

$$p_{o1} = \frac{p_1}{0.984} = \frac{345}{0.984} = 350.61\ kPa$$

From Fanno tables

At $M_1 = 0.15, \frac{\lambda_{av} L_{max}}{D} = 28.354,$

$$\frac{p_1}{p^*} = 7.319, \frac{T_1}{T^*} = 1.1945, \frac{p_{o1}}{p^*_{o_1}} = 3.928, \frac{V}{V^*} = 0.164$$

At $M_2 = 0.5, \frac{\lambda_{av} L_{max}}{D} = 1.069,$

$$\frac{p_2}{p^*} = 2.138, \frac{T_2}{T^*} = 1.143, \frac{p_{o2}}{p^*_{o_2}} = 1.34, \frac{V}{V^*} = 0.534$$

(b) $$\frac{\lambda_{av} L}{D} = \left(\frac{\lambda_{av} L_{max}}{D}\right)_{M_1} - \left(\frac{\lambda_{av} L_{max}}{D}\right)_{M_2}$$

$$= 28.354 - 1.069 = 27.285$$

$$L = \frac{27.285 \times 0.226}{0.02} = 308.32\ m$$ Ans

(c) $$T = \frac{(T_2/T^*)_{M_2}}{(T/T^*)_{M_1}} \times T_1 = \frac{1.143}{1.195} \times 311 = 297.467\ K$$ Ans

(d) $$p_2 = \frac{(p/p^*)_{M_2}}{(p/p^*)_{M_1}} \times p_1 = \frac{2.138}{7.319} \times 345 = 100.78\ kPa$$ Ans

$$p_{o2} = \frac{(p_o/p_o^*)_{M_2}}{(p_o/p_o^*)_{M_1}} \times p_{o_1} = \frac{1.34}{3.928} \times 350.61 = 119.61\ kPa$$ Ans

$$p_{o1} - p_{o2} = 350.61 - 119.61 = 231\ kPa$$ Ans

$$V_2 = \frac{(V/V^*)_{M_2}}{(V/V^*)_{M_1}} \times V_1 = \frac{0.534}{0.164} \times 53.024 = 172.65\ m/s$$

$$a_2 = \sqrt{k R T_2} = \sqrt{1.4 \times 287 \times 297.467} = 345.72\ m/s$$

$$M_2 = V_2/a_2 = \frac{172.65}{345.72} = 0.49939$$ Ans

This is approximately equal to 0.5.

5.10 What is $\lambda_{av} L^* /D$ for adiabatic flow in a pipe with friction (a) for the initial Mach number M_1 approaching zero, (b) for M_1 approaching 1, and (c) for very large values of M_1 (supersonic flow) ? Take k = 1.4.

Solution Using equation

$$\frac{\lambda_{av} L^*}{D} = \frac{\lambda_{av} L_{max}}{D} = \frac{1 - M_1^2}{k M_1^2} + \frac{k+1}{2k} \ln \frac{(k+1) M_1^2}{2\left(1 + \frac{k-1}{2} M_1^2\right)}$$

(a) The above equation becomes infinite as M_1 approaches zero, and thus $\lambda_{av} L^* /D$ approaches infinite. An indeterminate expression is obtained if M_1 =

0 is subtracted directly, but if the above equation is differentiated with respect to M_1 and $M_1 = 0$ is substituted, the limiting value is seen to be infinity.

(b) $\lambda_{av} L^*/D \rightarrow 0$ as $M_1 \rightarrow 1$ in above equation.

(c) $$Lt_{M_1 \to \infty} \frac{\frac{1}{M_1^2} - 1}{k} + \frac{k+1}{2k} \ln \left[\frac{k+1}{2\left(\frac{1}{M_1^2} + \frac{k-1}{2}\right)} \right]$$

$$= -\frac{1}{k} + \frac{k+1}{2k} \ln\left(\frac{k+1}{k-1}\right)$$

$$= -\frac{1}{1.4} + \frac{2.4}{2.8} \ln 6 = 0.822$$ Ans

5.11 Air flow in an insulated tube of 7.16 mm diameter by a vacuum pump. The air is drawn from a room where the absolute pressure is 101 kPa and the temperature is 296 K, through a smoothly contoured, congerging nozzle. At section one, where the nozzle joins the constant area tube, the static pressure is 98.5 kPa absolute. At section two, located some distance downstream in the constant area tube, the air temperature is 287 K. Determine the mass flow rate, the local isentropic stagnation pressure at section two, and the friction force on the duct wall between sections one and two.

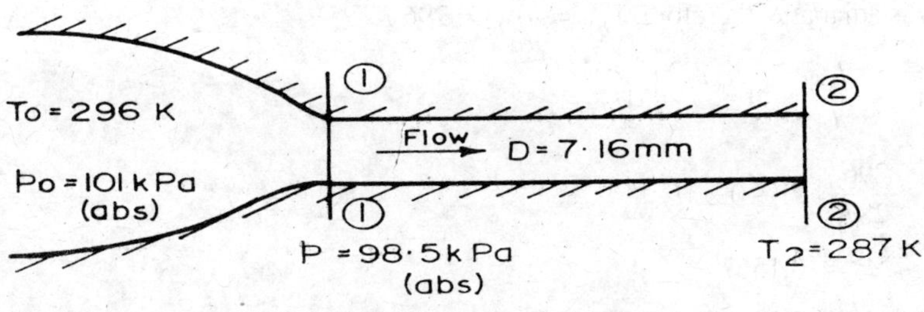

Fig 5.5 Insulated tube P 5.11

Solution

$$\frac{p_{o1}}{p_1} = \left(1 + \frac{k-1}{2} M_1^2\right)^{k/k-1}$$

or, $$1 + \frac{k-1}{2} M_1^2 = \left(\frac{101}{98.5}\right)^{0.286}$$

or, $$\frac{k-1}{2} M_1^2 = 1.007194 - 1$$

or, $$M_1^2 = \frac{2}{0.4} \times 7.194 \times 10^{-3}$$

or, $$M_1^2 = 0.03497$$

or, $$M_1 = 0.18965 \approx 0.19$$

$$\frac{T_{o1}}{T_1} = 1 + \frac{k-1}{2} M_1^2 = 1 + 0.2\,(0.19)^2 = 1.00722$$

or, $$T_1 = \frac{T_{o1}}{1.00722} = \frac{296}{1.00722} = 293.878$$

$$p_1 = \rho_1 R T_1$$

or, $$98.5 = \rho_1 \times 0.287 \times 293.878$$

or, $$\rho_1 = 1.1678\ kg/m^3$$

$$a_1 = \sqrt{k\,RT_1} = \sqrt{1.4 \times 287 \times 293.878} = 343.68\ m/s$$

$$V_1 = M_1\, a_1 = 0.19 \times 343.68 = 65.28\ m/s$$

$$A_1 = A = \frac{\pi}{4} D^2 = \frac{\pi}{4}\left(\frac{7.16}{1000}\right)^2 = 4.026 \times 10^{-5}\ m^2$$

$$\dot{m} = \rho_1\, A_1\, V_1 = 1.1678 \times 4.0264 \times 10^{-5} \times 65.28$$

$$= 3.0694 \times 10^{-3}\ kg/s$$ Ans

Flow is adiabatic, therefore, $T_{o1} = T_{o2} = 296\ K$ Ans

Then, $$\frac{T_{o2}}{T_2} = 1 + \frac{k-1}{2} M_2^2$$

$$\frac{296}{287} = 1 + 0.2\, M_2^2$$

$$M_2^2 = 0.15679$$

$$M_2 = 0.39597 \approx 0.396$$ Ans

$$a_2 = \sqrt{k\,RT_2} = \sqrt{1.4 \times 287 \times 287} = 339.583\ m/s$$

$$V_2 = M_2\, a_2 = 0.396 \times 339.583 = 134.47\ m/s$$

From the continuity equation

$$\rho_2\, A_2\, V_2 = \rho_1\, A_1\, V_1$$ $\{Here\ A_1 = A_2\}$

or, $$\rho_2 = \frac{\rho_1\, V_1}{V_2} = \frac{1.1678 \times 65.28}{134.47} = 0.567\ kg/m^3$$ Ans

$p_2 = \rho_2 RT_2 = 0.567 \times 287 \times 287$

$= 46.695 \approx 46.7\ kPa$ Ans

$$p_{o2} = p_2 \left(1 + \frac{k-1}{2} M_2^2\right)^{k/(k-1)}$$

$= 46.7\ [1 + 0.2\ (0.396)^2]^{3.5}$

$= 46.7 \times 1.114 = 52.03$ kPa

The friction force may be calculated by applying the momentum theorem

$$-F_f + p_1 A - p_2 A = \dot{m} \quad (V_2 - V_1)$$

$$-F_f = (p_2 - p_1) A + \dot{m} \quad (V_2 - V_1)$$

$= (46.7 - 98.5) \times 4.0264 \times 10^{-5}$

$+ 3.0694 \times 10^{-3}\ (134.47 - 65.28)$

$= -2.08567 + 0.21237 = -1.8733$ N

$F_f = 1.8733$ N (to the left as assumed) This is the force exerted on the control volume by the duct wall. The force off the fluid on the duct is

$R_x = F_f = 1.8733\ N$ (*to the right*) Ans

5.12 Estimate the maximum flow rate of air through the passage shown in Fig. Assuming friction factor of the duct to be 0.02. What range of back pressure will this flow rate be achieved ?

Solution

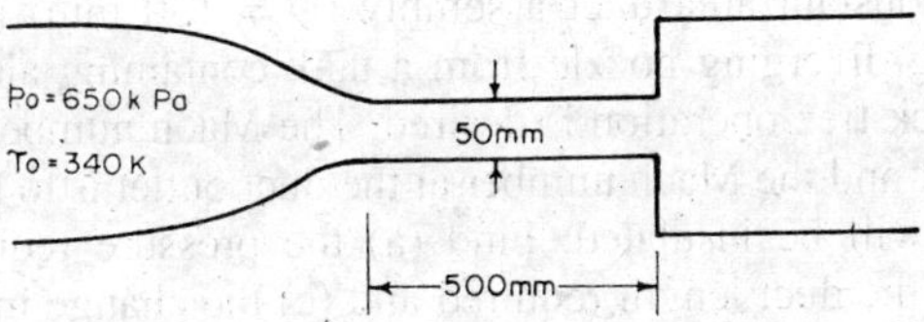

Fig 5.6 Air flow through passage P 5.12

$$A = \frac{\pi}{4} D^2 = \frac{\pi}{4} (0.05)^2 = 1.963 \times 10^{-3}\ m^2$$

For the Fig 5.6, the condition at the end of the duct is Mach number should be unity.

Using equation

$$\frac{\lambda_{av} L_{max}}{D} = \frac{0.02 \times 500}{50} = 0.2$$

From Fanno flow tables

At $\frac{\lambda_{av} L_{max}}{D} = 0.2, \quad M = 0.7, \frac{T}{T^*} = 1.093, \frac{p}{p^*} = 1.493$

From isentropic tables

At $M = 0.7, \quad p/p_o = 0.721$

$$p = 0.721 \times p_o = 0.721 \times 650 = 468.65 \; kPa$$

Also, $\dot{m} = A \sqrt{\frac{k}{R}} \times \frac{P}{\sqrt{T_o}} . M \sqrt{1 + \frac{k-1}{2} M^2}$

$$\dot{m} = 1.963 \times 10^{-3} \sqrt{\frac{1.4}{287}}$$

$$\times \frac{468.65 \times 10^3}{\sqrt{340}} \times 0.7 \times \sqrt{1 + 0.2\,(0.7)^2}$$

$$\dot{m} = 2.55 \; kg/s$$

Also, $\frac{p^*}{p_o} = \frac{p^*/p}{p_o/p}$

$$= \frac{1}{1.493} \times 0.721 = 0.483$$

$$p^* = 0.483 \times p_o = 0.483 \times 650 = 313.898 = 313.9 \; kPa$$

Therefore, $p^* \leq 313.9$ kPa Ans

5.13 A smooth constant area duct assembly (D = 150 mm) is to be fed by a converging diverging nozzle from a tank containing air at 295 K and 1 MPa. Shock free operation is desired. The Mach number at the duct inlet is to be 2.1 and the Mach number at the duct outlet is to be 1.4. The entire assembly will be insulated. Find (a) the pressure required at the duct outlet, (b) the duct length required and (c) the change in specific entropy along the duct. Take $\lambda_{av} = 0.007$.

Solution From isentropic tables

At $M_1 = 2.1, \quad \frac{T_1}{T_o} = 0.5314, \quad \therefore \; T_1 = 156.7 \; K$

$$\frac{p_1}{p_o} = 0.1094, \; p_1 = 109.4 \; kPa$$

$$a_1 = \sqrt{k\,RT_1} = \sqrt{1.4 \times 287 \times 156.76} = 250.97 \; m/s$$

$$V_1 = M_1 \, a_1 = 2.1 \times 250.97 = 527.0429 \; m/s$$

From Fanno tables

At $M_1 = 2.1,\ p_{o1}/p_o^* = 1.837,\ p_o^* = 544.36\ kPa$

$$p_1/p^* = 0.3802,\ p^* = 287.74\ kPa$$

and at $M_2 = 1.4,\ p_2/p^* = 0.6632,\ p_2 = 190.83\ kPa$ Ans

$$p_{o2}/p_o^* = 1.115,\ p_{o2} = 606.96\ kPa$$ Ans

At $M_1 = 2.1, \left(\frac{\lambda_{av} L_{max}}{D}\right)_{M_1} = 0.3339$

and at $M_2 = 1.4, \left(\frac{\lambda_{av} L_{max}}{D}\right)_{M_2} = 0.09074$

Therefore,

$$\frac{\lambda_{av} L}{D} = \left(\frac{\lambda_{av} L_{max}}{D}\right)_{M_1} - \left(\frac{\lambda_{av} L_{max}}{D}\right)_{M_2}$$

$$= 0.3339 - 0.09974 = 0.23416$$

$$L = 0.23416 \times \frac{150}{1000} \times \frac{1}{0.007} = 5.01\ m$$ Ans

$$s_2 - s_1 = -R \ln \frac{p_{o2}}{p_{o1}}$$

$$= -287 \ln \frac{606.96}{1000}$$

$$= -287 \ln (0.60696) = 143\ \text{J/kg K}$$ Ans

5.14 Air from a converging nozzle to a tube of 40 mm diameter and length 500 mm. The stagnation condition at inlet are 1 MPa and 333 K and flow in nozzle is isentropic and in the tube is adiabatic. The average friction factor $\lambda_{av} = 0.016$. Determine the mass flow rate when the exit pressure of the tube is 340 kPa and 680 kPa respectively.

Solution

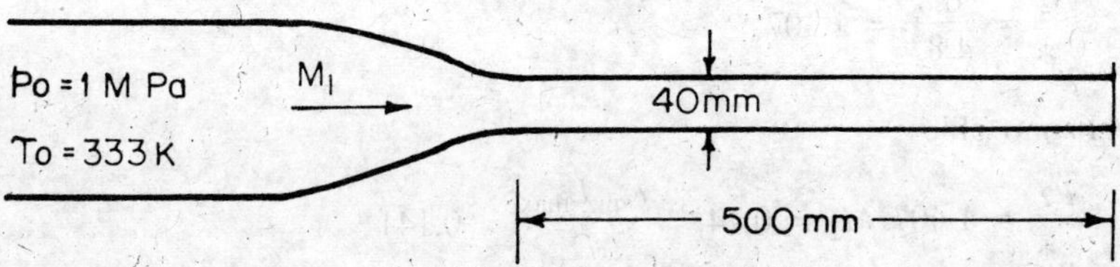

Fig 5.7 Converging nozzle P 5.15

$$A = \frac{\pi}{4}(0.04)^2 = 1.256 \times 10^{-3}\ m^2$$

We know that $\frac{\lambda_{av} L_{max}}{D} = \frac{0.016 \times 500}{40} = 0.2$

From Fanno tables

At $\frac{\lambda_{av} L_{max}}{D} = 0.2,\ M_1 = 0.7,\ \frac{p_1}{p^*} = 1.493$

Since flow in the nozzle is isentropic

From isentropic tables

At $M_1 = 0.7,\ \frac{p_1}{p_o} = 0.721$

$$p_1 = 0.721 \times p_o = 0.721 \times 1 = 0.721\ MPa = 7.21\ bar$$

and $p^* = \frac{7.21}{1.493} = 4.83\ bar$

As the pressure pe = 3.4 bar is less than the pressure p^* = 4.83 bar, we may conclude that maximum length so assumed is correct. There cannot be any decrease in Mach numbers at exit of the nozzle. The tube is choked adiabatically.

$$\dot{m} = A \frac{p_o}{\sqrt{T_o}} \sqrt{\frac{k}{R}} \frac{M}{\left(1 + \frac{k-1}{2} M^2\right)^{(k+1)/2(k-1)}}$$

$$\dot{m} = 1.256 \times 10^{-3} \times \frac{10 \times 10^5}{\sqrt{333}}$$

$$\times \sqrt{\frac{1.4}{287}} \times \frac{0.7}{[1 + 0.2(0.7)^2]^3} = 2.542\ kg/s.$$

(b) When the exit pressure is 6.8 bar, choked condition cannot be assumed and length is not the maximum length for M_1. Thus, We have

$$p^* = 4.83\ bar,\ p_2 = 6.8\ bar$$

and $\frac{p_2}{p^*} = \frac{6.8}{4.83} = 1.407$

From Fanno tables

At $\frac{p_2}{p^*} = 1.407,\ M_2 = 0.74,\ \frac{\lambda_{av} L_{max}}{D} = 0.141$

$$\text{Also, } \frac{\lambda_{av} L}{D} = \left(\frac{\lambda_{av} L_{max}}{D}\right)_{M_1} - \left(\frac{\lambda_{av} L_{max}}{D}\right)_{M_2}$$

$$0.2 = \left(\frac{\lambda_{av} L_{max}}{D}\right)_{M_1} - 0.141$$

$$\left(\frac{\lambda_{av} L_{max}}{D}\right)_{M_1} = 0.2 + 0.141 = 0.341$$

From Fanno tables

$$\text{At} \quad \left(\frac{\lambda_{av} L_{max}}{D}\right)_{M_1} = 0.341; \quad M_1 \approx 0.64$$

Thus when $L > L_{max}$ initial Mach number is reduced

Mass flow rate

$$\dot{m} = A \frac{p_o}{\sqrt{T_o}} \times \sqrt{\frac{k}{R}} \frac{M_1}{\left(1 + \frac{k-1}{2} M_1^2\right)^{(k+1)/2(k-1)}}$$

$$= 1.256 \times 10^{-3} \times \frac{10 \times 10^5}{\sqrt{333}} \times \sqrt{\frac{1.4}{287}} \times \frac{0.64}{[1 + 0.2\,(0.64)^2]^3}$$

$$= 2.429\ kg/s$$ Ans

5.15 A convergent-divergent nozzle supplies air to a constant diameter duct. The nozzle diameters at the throat and exit are 150 mm and 300 mm respectively. Duct diameter is also 400 mm. The stagnation pressure at the nozzle entry is 10 bar and that at section one, 2.5 D downstream of the nozzle exit section is 40 kPa. The flow at the duct exit section two is sonic. For a friction factor of 0.012 determine

(a) Mach number at section one and nozzle exit

(b) Static pressure at section two and nozzle exit, and

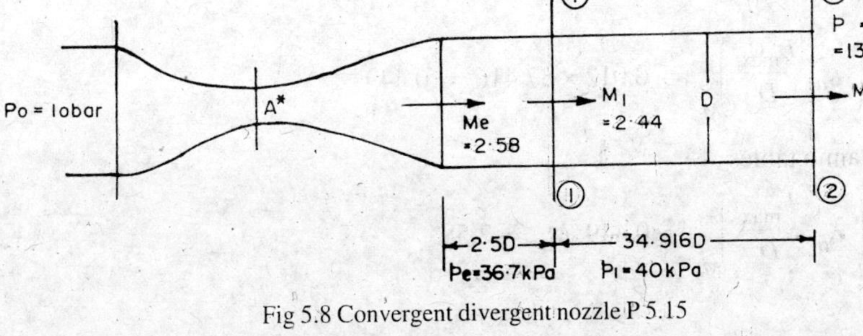

Fig 5.8 Convergent divergent nozzle P 5.15

(c) total duct length from the nozzle to the exit.

Assume isentropic flow up to nozzle throat and adiabatic further downstream up to the duct exit.

Solution

$$\frac{A_1}{A^*} = \left(\frac{D_1}{D^*}\right)^2 = \left(\frac{300}{150}\right)^2 = 4$$

$$\frac{p_1}{p_o} = \frac{0.4}{10} = 0.04$$

Therefore at point one

$$\frac{A_1 p_1}{A^* p_o} = 4 \times 0.04 = 0.16$$

From isentropic tables

At $\frac{A_1 p_1}{A^* p_o} = 0.16, \quad M_1 = 2.44$ Ans

From Fanno tables

At $M_1 = 2.44, \quad \frac{p_1}{p^*} = 0.303, \frac{\lambda_{av} L_{max}}{D} = 0.419$

Therefore, $p^* = \frac{p_1}{0.303} = \frac{40}{0.303} = 132\ kPa$

$$\left(\frac{L_{max}}{D}\right)_{M_1} = \frac{0.419}{0.012} = 34.916$$

At point e,

$$\left(\frac{L_{max}}{D}\right)_{M_e} = 34.916 + 2.5 = 37.416$$

$$\left(\lambda_{av}\frac{L_{max}}{D}\right)_{M_e} = 0.012 \times 37.416 = 0.449$$

From Fanno tables

At $\left(\lambda_{av}\frac{L_{max}}{D}\right)_{M_e} = 0.449, M_e = 2.58$ Ans

$$\frac{p_e}{p^*} = 0.278$$

Therefore, $p_e = 0.278 \times 132 = 36.696\ kPa$ Ans

From isentropic tables,

At $\frac{p_e}{p_o} = \frac{36.696}{1000} = 0.036696$

$M_e = 2.8$ Ans

5.16 A convergent divergent nozzle is provided with a pipe of constant cross section at its exit; the exit diameter of the nozzle and that of the pipe is 400 mm. The average coefficient of friction for the pipe is 0.01. Stagnation pressure and temperature of air at the nozzle entry are 1.2 MPa and 600 K. The flow is isentropic in the nozzle and adiabatic in the pipe. The Mach numbers at the entry and exit of the ppe are 1.8 and 1 respectively. Determine,

(a) the length of the pipe.

(b) the diameter of the nozzle throat, and

(c) pressure and temperature at the pipe exit.

Solution

$$A_1 = \frac{\pi}{4} D^2 = \frac{\pi}{4} (0.4)^2 = 0.1256\ m^2$$

From isentropic flow tables

At $M_1 = 1.8$

$$\frac{p_1}{p_{o1}} = 0.174, \frac{T_1}{T_{o1}} = 0.607, \frac{A_1}{A^*} = 1.439$$

$$A^* = \frac{A_1}{1.439} = \frac{0.1256}{1.439} = 0.087327\ m^2$$

$$T_1 = 0.607 \times 600 = 364.2\ K$$

$$p_1 = 0.174 \times 1.2 = 0.2088\ M\ Pa$$

$$\frac{\pi}{4} D^{*2} = 0.087327$$

$\therefore$ $D^* = 0.3334\ m = 333.4\ mm$ Ans

From Fanno tables

At $M_1 = 1.8, \dfrac{p_1}{p^*} = 0.474, \dfrac{T_1}{T^*} = 0.728, \dfrac{\lambda_{av} L_{max}}{D} = 0.242$

or, $\dfrac{0.01\,L_{max}}{0.4} = 0.242$

or, $L_{max} = 9.68\ m$ Ans

Since the flow is sonic at the pipe exit therefore length of the pipe is 9.68 m and pressure and temperature are p^* and T^* respectively.

$$p^* = \frac{p_1}{0.474} = \frac{0.2088}{0.474} = 0.44\ MPa$$

$$T^* = \frac{T_1}{0.728} = \frac{364.2}{0.728} = 500.274725\ K \qquad \text{Ans}$$

5.17 Air flow is induced in a smooth insulated tube of 8 mm diameter by a vacuum pump. The air is drawn from a room where the absolute pressure is 760 mm of mercury and the temperature is 296 K through a smoothly contoured, converging nozzle. At section one where the nozzle joins the constant area tube, the static gauge pressure is – 21.5 mm of mercury. At section two, located some distance downstream in the constant area tube, the static pressur is – 412 mm of mercury gauge. The duct walls are smooth; the average friction factor λ_{av} may be taken as the value at section one. Determine the length of duct required for choking from section one, the Mach number at section one, the Mach number at section two, and the duct length, L_{12} between sections one and two. Take $\mu = 1.9 \times 10^{-5}\ Ns/m^2$, $\lambda_{av} = 0.0232$.

Solution

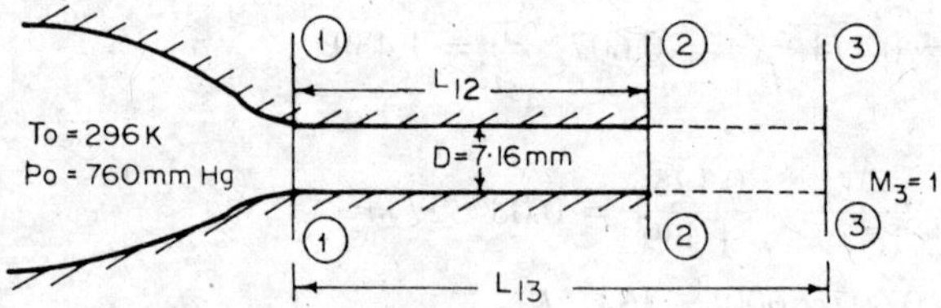

Fig 5.9 Smooth insulated tube P 5.17

$$p_{o1} = 760\ mm\ of\ H_g = 1.01325\ bar,$$

$$p_1 = \frac{760 - 21.5}{750} = 0.985\ bar$$

$$\frac{p_{o1}}{p_1} = \left(1 + \frac{k-1}{2} M_1^2\right)^{k/(k-1)} \quad \text{or,} \quad \frac{k-1}{2} M_1^2 = \left(\frac{p_{o1}}{p^1}\right)^{(k-1)/k} - 1$$

$$0.2\,M_1^2 = \left(\frac{1.01325}{0.985}\right)^{0.286} - 1$$

$$= 1.0081199 - 1$$

$$= 8.1199 \times 10^{-3}$$

$$M_1^2 = 0.0405995$$

$$M_1 = 0.2$$

$$\frac{T_{o1}}{T_1} = 1 + \frac{k-1}{2} M_1^2 = 1 + 0.2\,(0.2)^2 = 1.008$$

$$T_1 = \frac{T_{o1}}{1.008} = \frac{296}{1.008} = 293.65\ K$$

$$a_1 = \sqrt{k\,RT_1} = \sqrt{1.4 \times 287 \times 293.65} = 343.495$$

$$V_1 = M_1\,a_1 = 0.2 \times 343.495 = 68.7\ m/s$$

$$p_1 = \rho_1\,RT_1$$

$$98.5 = \rho_1 \times 0.287 \times 293.65$$

$$\rho_1 = 1.16875\ kg/m^3$$

$$R_{e_1} = \frac{\rho_1\,V_1\,D}{\mu_1} = \frac{1.16875 \times 68.7 \times 8}{1.9 \times 10^{-5} \times 1000} = 33807.84$$

Taking $\lambda_{av} = \lambda_1$

From Fanno tables

At $M_1 = 0.2,\ \frac{p}{p^*} = 5.455,\ \frac{\lambda_{av}\,L_{max}}{D} = 14.533$

$$L_{13} = (L_{max})_{M_1} = \left(\frac{\lambda_{av}\,L_{max}}{D}\right)_{M_1} \times \frac{D}{\lambda_1}$$

$$= 14.533 \times \frac{8}{1000} \times \frac{1}{0.0232} \approx 5\ m$$ Ans

$$p_2 = (760 - 412)/750 = 0.464\ bar$$

Since p^* is constant for Fanno line flow, condition at point 2 can be determined from the pressure ratio, $(p/p^*)_{M_2}$. Thus

$$\left(\frac{p}{p^*}\right)_{M_2} = \frac{p_2}{p^*} = \frac{p_2}{p_1} \times \frac{p_1}{p^*} = \frac{p_2}{p_1}\left(\frac{p}{p^*}\right)_{M_1}$$

$$= \frac{46.4}{98.5} \times 5.455 = 2.5696$$

From Fanno tables

At $\quad p_2/p^* = 2.5695, M_2 = 0.42.$ Ans

At $\quad M_2 = 0.42, \frac{\lambda_{av} L_{max}}{D} = 1.974$

$$\text{Thus, } L_{23} = \left(L_{max}\right)_{M_2} = \left(\frac{\lambda_{av} L_{max}}{D}\right) \times \frac{D}{\lambda_1}$$

$$= 1.974 \times \frac{8}{1000} \times \frac{1}{0.0232} = 0.68\ m$$

Finally, $L_{12} = L_{13} - L_{23} = 5 - 0.68 = 4.32\ m$ Ans

EXERCISES

A. THEORY

5.1 Write down the assumption used for defining the Fanno flow.

5.2 Define Fanno flow and write the assumptions for defining the Fanno flow.

5.3 Point wise give the important features of Fanno curve.

5.4 Give three examples of Fanno flow in thermal systems; give reasons justifying Fanno flow in each of these cases.

5.5 Derive an equation describing a Fanno curve. Show three Fanno curves on the T–s or h–s coordinates at three mass flow densities.

5.6 Show that the upper and lower branches of a Fanno curve represent subsonic and supersonic flows respectively. Prove that at the maximum entropy point Mach number is unity and all processes approach this point.

How would the state of a gas in a flow change from the supersonic to subsonic branch ?

5.7 Draw three T–s diagrams for subsonic, supersonic and normal shock occurance in Fanno flow and explain the behaviour of the curves.

5.8 Fanno flow behaviour can be studied very easily by using the different theromodynamic parameters. Prepare the table and explain.

5.9 Explain the effects of wall friction on fluid properties.

5.10 Develop the following relations for Fanno Flow.

(a) $\quad \frac{dT}{T} + \left(\frac{k-1}{2}\right) M^2 \frac{dV^2}{V^2} = 0$

(b) $$\frac{d\rho}{\rho} + \frac{1}{2}\left(\frac{dV^2}{V^2}\right) = 0$$

(c) $$\frac{dV^2}{V^2} = \frac{dp}{p}\left[\frac{2}{1 + (k-1)\,M^2}\right]$$

(d) $$\frac{dp}{p} + \frac{kM^2}{2}\left(\frac{\lambda\, dx}{D}\right) + \frac{kM^2}{2}\left(\frac{dV^2}{V^2}\right) = 0$$

(e) $$\frac{dp_o}{p_o} = \frac{dp}{p} + \frac{kM^2}{2\left(1 + \frac{k-1}{2} M^2\right)}\left(\frac{dM^2}{M^2}\right)$$

(f) $$\frac{dF}{F} = \frac{kM^2}{1 + kM^2}\left(\frac{dM^2}{M^2}\right) + \frac{dp}{p}$$

(g) $$\frac{\rho^*}{\rho} = \frac{V^*}{V} = \sqrt{\left[\frac{(k+1)\,M^2}{2 + (k-1)M^2}\right]}$$

(h) $$\frac{p}{p^*} = \frac{1}{M}\sqrt{\left(\frac{T}{T^*}\right)} = \frac{1}{M}\sqrt{\frac{k+1}{2 + (k-1)\,M^2}}$$

(i) $$\frac{p_o}{p_o^*} = \left(\frac{1}{M}\right)^{2k/(k-1)}\left(\frac{p^*}{p}\right)^{(k+1)/(k-1)}$$

$$= \frac{1}{M}\left[\frac{2 + (k-1)M^2}{k+1}\right]^{(k+1)/2\,(k-1)}$$

(j) $$\frac{F}{F^*} = \frac{1 + kM^2}{M\left[\,(k+1)\left\{2 + (k-1)\,M^2\right\}\right]^{1/2}}$$

(k) $$s^* - s = R \ln \frac{1}{M}\left[\frac{2 + (k-1)\,M^2}{(k+1)}\right]^{(k+1)/2\,(k-1)}$$

(l) $$\left(\frac{\lambda_{av}\, L_{\max}}{D}\right)_M = \frac{1 - M^2}{kM^2} + \frac{k+1}{2k}\ln\frac{(k+1)\,M^2}{2 + (k-1)\,M^2}$$

(m) $$\frac{p_2}{p_1} = \frac{(p/p^*)_{M_2}}{(p/p^*)_{M_1}} = \frac{M_1}{M_2}\sqrt{\frac{1 + \frac{(k-1)}{2} M_1^2}{1 + \frac{(k-1)}{2} M_2^2}}$$

(n) $$\frac{\lambda_{av}\, L}{D} = \left(\frac{\lambda_{av}\, L_{\max}}{D}\right)_{M_1} - \left(\frac{\lambda_{av}\, L_{\max}}{D}\right)_{M_2}$$

B. UNSOLVED PROBLEMS

5.1 The Mach number for air flow in a pipe increases from 0.1 to 0.6 because of pipe friction. The presssure at the section where $M_1 = 0.1$ is $p_1 = 550$ kPa. What is the pressure at the section where $M_2 = 0.6$ when the flow is adiabatic. [88.6 kPa]

5.2 Methane flows through a 200 mm insulated pipe at a rate of 11 kg/s. At inlet the pressure is 700 kPa and the temperature is 50°C. Calculate (a) the initial Mach number, (b) the Mach number at section where the temperature is 40°C, and (c) the pressure at the section where the temperature is 40°C. [$M_1 = 0.179$, $M_2 = 0.488$, 252 kPa]

5.3 Natural gas is pumped through a 500 mm commercial steel pipe at a rate of 150 kg/s compressors raise the pressure from 3100 to 5250 kPa, and the gas is transmitted to the next compressor station. Assume the gas has properties of methane and is at 40°C. How far apart are the compressor stations ? Take $\lambda_{av} = 0.0315$. [6990 m]

5.4 Air is flowing through a well insulated, constant area channel. At a section where the speed is 144 m/s, t = 50°C and p = 600 kPa. Determine (a) the temperature and stagnation pressure at the downstream section where the density is 3.05 kg/m^3 and (b) the entropy increase. [287 K, 423 kPa, 132 J/kg K]

5.5 Air at a pressure of 500 kPa and temperature 300 K contained in a pressure vessel which has a wall thickness of 6 mm is found to be leaking at a rate of 2×10^{-6} kg/s to atmosphere. Assuming that there is a small circular hole of length equal to wall thickness, determine its diameter. It may be assumed that flow is choked at outlet from the hole and that flow is isentropic upto the entrance of the hole. Take $\frac{\lambda_{av}}{D} = 0.188$. [D = 0.065 mm]

5.6 Air flows adiabatically with friction through a duct of 0.3 m square cross section. At one section, the pressure, temperature and speed are 60 kPa, 50°C and 180 m/s. At the exit from the duct, the pressure and temperature are 31.7 kPa and 19°C. Calculate the exit Mach number and the stagnation temperature midway between the first section and the exit. [$M_e = 0.897$, $T_{o2} = 339\ K$]

5.7 Air flows steadily and adiabatically through a constant area duct. The density at the inlet is 3.51 kg/m^3. If the exit Mach number is unity and the exit temperature and pressure are 227°C and 110 kPa, respectively, determine the inlet Mach number and the entropy change from inlet to exit. [M_1 = 0.2, 311 J/kg K]

5.8 The jet pipe on a jet installation is 10 m long, 0.5 m in diameter and terminates in a short converging nozzle with an exit diameter of 0.46 m. The stagnation temperature and pressure at entry to the pipe is 800°C and

300 kPa respectively. The engine is run with aircraft stationary on the ground and atmospheric pressure is 101.3 kPa. Calculate net thrust of the engine (a) When the friction in the pipe is neglected. (b) when the friction coefficient $\tau_{\omega}/\rho\, V^2/2$ in the jet pipe is 0.01. Assume K = 1.33, c_p = 1.15 kJ/kg K. [(a) 45.6 kN, (b) 36.86 kN]

5.9 Air flows through an insulated duct with constant area of 0.03 m^2. At section one, the static temperature, static pressure, and Mach number are 277 K, 690 kPa, and 0.6. At section two downstream, the temperature is 260 k. Calculate M_2, and p_2. Evaluate the frictional force exerted by the air on the duct wall. [M_2 = 0.844, p_2 = 475 kPa, F = 2650 N (to the right)].

5.10 Measurements are made of compressible flow in a long smooth 7.16 mm internal diameter tube. Air is drawn from the surroundings (20°C and 101 kPa) by a vacuum pump downstream. Pressure readings along the tube become steady when the downstream presssure is reduced to 626 mm H_g (vacuum). For these conditions, determine (a) the maximum mass flow rate possible through the tube, (b) the stagnation pressure of the air leaving the tube, and (c) the entropy change of the air in the tube. $\dot{m}$ = 0.00321 kg/s p_{o2} = 33.8 kPa, $s_2 - s_1$ = 314 J/kg K].

5.11 A converging diverging nozzle discharges air into an insulated pipe with area A = 650 mm^2. At the pipe inlet, p_1 = 128 kPa t_1 = 39°C, and M_1 = 2. For shockless flow to a Mach number of unity at the pipe exit, calculate the exit temperature, the net force of the fluid on the pipe, and the entropy change. [T_e = 468 K, 600 N (to right) 149 J/kg K]

5.12 Air flows steadily and adiabatically in a horizontal tube of 50 mm diameter. Measurements at section one show that the temperature and pressure are 53°C and 340 KPa. At section two downstream, the temperature is 114°C and the flow speed is 591 m/s. Find (a) the flow speed, and Mach number at section one and (b) the magnitude and direction of the friction force exerted by the air on the duct wall. [M_1 = 1.9, V_1 = 688 m/s, 220 N (to the right)].

5.13 A stream of air flows in a insulated tube of constant area having a cross sectional area of 0.9 m^2. At a section one, the pressure is 0.6 bar, temperature is 22°C and mass velocity is 150 kg/m^2s. The pressure in space in which tube exhausts is so low that choking condition prevails. Calculate (a) Mach number at section one (b) total force exerted in axial direction which must be exerted to hold duct stationary. [M_1 = 0.6, – 1462.5 N]

5.14 Using coordinates T/T^* and $(s - s^*)/c_p$, where s^* is the entropy at the condition where M = 1, plot the Fanno line for air flow (k = 1.4) for Mach numbers in the range 0.1 < M < 3.

5.15 Beginning with inlet conditions of problem 5.11, and using coordinates T/T_0 and $(s - s^*)/c_p$, plot the supersonic and subsonic brances of the Fanno line for the flow.

5.16 For the conditions of problem 5.11, determine the duct length. Assume that the duct is circular and made from commercial steel. Plot the variations of pressure and Mach number versus distance along the duct.

6

NORMAL SHOCK

6.1 INTRODUCTION

A normal shock waves is one in which the plane of the shock wave is at right angles to the flow streamlines.

Normal shock waves can occur in supersonic flows through converging diverging and constant area ducts.

Normal shock waves decelerations from a supersonic flow to subsonic flow, a pressure rise, and an increase of entropy.

For practically all engineering calculations we are concerned primarily with the net changes in fluid propeties across the shock, and have no interest in the complex viscous and heat conduction phenomena in the interior of the shock wave. It is profitable, therefore, to ignore the interior details of the shock wave, and to employ a simplified model, namely, a pure discontinuity.

We therefore focus our attention on the relation between the stream properties on the two sides of such a discontinuity, and on the question of whether all such discontinuities are possible. Although experiments show that these discontinuities are often oblique to the direction of flow.

6.2. Governing relations of the normal shock

Let us consider a control volume as in Fig 6.1, specified, so as to include the normal shock and a small quantity of fluid on each side of the discontinuity.

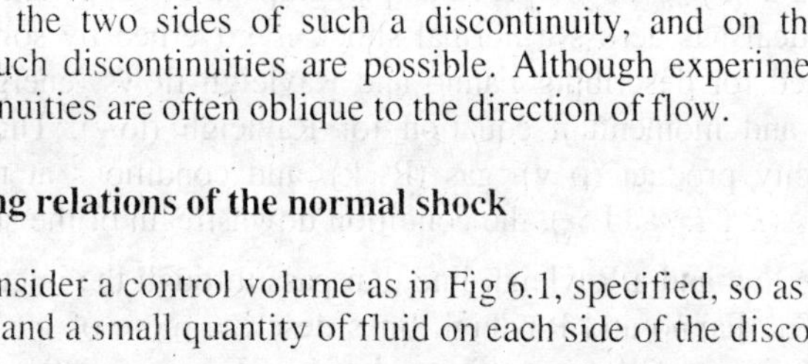

Fig. 6.1. Control volume including a normal discontinuity

It is common practice to designate conditions upstream of a shock with a subscript x and those downstream by the subscript y and these are indicated in Fig 6.1

The continuity equation is

$$\rho_x A V_x = \rho_y A V_y$$

or, $$\rho_x V_x = \rho_y V_y = constant \tag{6.1}$$

The energy equation is

$$h_x + \frac{V_x^2}{2} = h_y + \frac{V_y^2}{2} = h_o \tag{6.2}$$

Which indicates that the stagnation enthalpy is the sam on both sides of the shock.

The momentum equation may be written for the control volume of Fig 6.1, thus giving a third basic equation

$$p_x A - p_y A = \rho_x A V_x\left(V_y - V_x\right) = \rho_y A V_y\left(V_y - V_x\right)$$

or, $$p_x + \rho_x V_x^2 = p_y + \rho_y V_y^2 \tag{6.3}$$

which indicates tht the thrust function is the same on both sides of a normal shock.

The equation of state of the fluid may be writtne implicitly in the form

$$h = f(s, \rho)$$
$$s = f(p, \rho) \tag{6.4}$$

From the analyses of the previous paragraphs, it is apparent that the steady flow of an ideal gas across a normal shock is governed by some of the same equations used for describing Fanno and Rayleigh flows (energy equation for Fanno flow and momentum equation for Rayleigh flow). Thus, for a given density-velocity product (ρ v), gas (R, k), and conditions at the inlet of the normal shock (P_x, T_x, and S_x), the condition downstream of the shock will be on both a Fanno line and a Rayleigh line that pass through the inlet state as shown in Fig 6.2. The Fanno and Rayleigh lines describe more of the flow field than just in the vicinity of the normal shock when Fanno and Rayleigh flows are actually involved (Solid lines in Fig 6.3(a) and 6.3(b)). Otherwise, these lines (dashed lines in Fig 6.3(a), 6.3(b) and 6.3(c)) are useful mainly as a way to better visualize how the governing equations combine to yield a solution to the normal shock flow problem.

For Fig 6.2 (a) The normal shock in Fanno line

(b) The normal shock in Rayleigh line

(c) The normal shock in a frictionless and adiabatic flow.

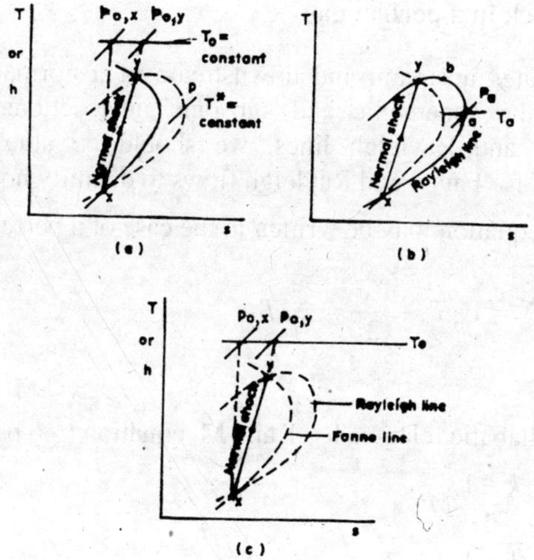

Fig. 6.2. The relationship between a normal shock and Fanno and Rayleigh lines

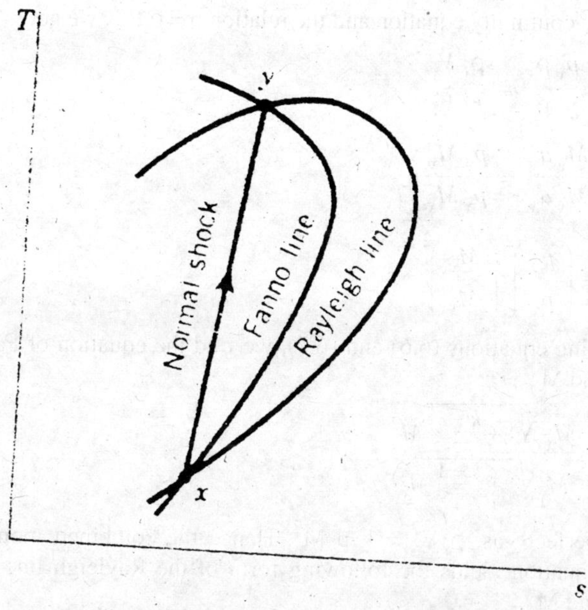

Fig. 6.3. Normal shock for Fanno and Rayleigh flow

According to the second law of themodynamics, the entropy can not decrease during an adiabatic change. Hence, only the change from x to y is possible. In other words, the normal discontinuity always involves a change from supersonic to subsonic speed with the reverse. This conclusion will, in the following article, be demonstrated rigorously for a perfect gas.

6.3. Normal shock in a perfect gas

Since the states upstream and downstream of a normal shock waves are represented by the supersonic and subsonic intersections of actual and/or imagined Fanno and Rayleigh lines, we should be able to use equations developed earlier for Fanno and Rayleigh flows to quantity normal shock flow.

The energy equation may be written in the case of a perfect gas as

$$c_p T_x + \frac{V_x^2}{2} = c_p T_y + \frac{V_y^2}{2} = c_p T_o \quad (6.5\text{ a})$$

or, $$T_{ox} = T_{oy} \quad (6.5\text{ b})$$

Using the adiabatic relation T_o, T and M, equation (6.6 b) takes the form

$$\frac{T_y}{T_x} = \frac{1 + \frac{k-1}{2} M_x^2}{1 + \frac{k-1}{2} M_y^2} \quad (6.6)$$

From the continuity equation and the relation $p = \rho RT$, we get

$$\frac{T_y}{T_x} = \frac{p_y \rho_x}{p_x \rho_y} = \frac{\rho_y V_y}{p_x V_x}$$

$$= \frac{p_y M_y a_y}{p_x M_x a_x} = \frac{p_y M_y \sqrt{T_y}}{p_x M_x \sqrt{T_x}}$$

or, $$\frac{T_y}{T_x} = \left(\frac{p_y}{p_x}\right)^2 \left(\frac{M_y}{M_x}\right)^2 \quad (6.7)$$

Combining equations (6.6) and (6.7), we find the equation of Fanno line in terms of p and M

$$\frac{p_y}{p_x} = \frac{M_x \sqrt{1 + \frac{k-1}{2} M_x^2}}{M_y \sqrt{1 + \frac{k-1}{2} M_y^2}} \quad (6.8)$$

For a perfect gas, $\rho V^2 = k p M^2$. Hence the combined momentum and continuity equation yields the following form of the Rayleigh line equation in terms of p and M

$$\frac{p_y}{p_x} = \frac{1 + k M_x^2}{1 + k M_y^2} \quad (6.9)$$

Finally, after eliminating p_y/p_x from equations (6.8) and (6.9), we get a relation between M_x and M_y.

$$\frac{M_x\sqrt{1+\frac{k-1}{2}M_x^2}}{1+kM_x^2} = \frac{M_y\sqrt{1+\frac{k-1}{2}M_y^2}}{1+kM_y^2} \tag{6.10}$$

Solving the above equation (6.10) explicitly for M_y we get the two solutions

$$M_y = M_x \tag{6.11}$$

and
$$M_y^2 = \frac{M_x^2 + \frac{2}{k-1}}{\frac{2k}{k-1}M_x^2 - 1} \tag{6.12}$$

This equation expresses the Mach number on the downstream side of the discontinuity in terms of the Mach number on the upstream side. If a normal shock exists in a fluid moving with a certain Mach number M_x, the Mach number following the shock must be as given by equation (6.12); regardless of other flow variables or fluid properties.

6.4 Prandtl-Meyer relation

Applying the adiabatic energy equation (2.35) to the flow before and after the shock wave we get

$$\frac{a_x^2}{k-1} + \frac{V_x^2}{2} = \frac{a_y^2}{k-1} + \frac{V_y^2}{2} = \left(\frac{k+1}{k-1}\right)\frac{a_x^2}{2}$$

From part of the above equation gives

$$a_x^2 = \frac{k+1}{2}a^{*2} - \frac{k-1}{2}V_x^2$$

or,
$$\frac{a_x^2}{V_x} = \frac{k+1}{2}\left(\frac{a^{*2}}{V_x}\right) - \frac{k-1}{2}V_x \tag{6.13}$$

Similarly the other part of the equation gives

$$\frac{a_y^2}{V_y} = \frac{k+1}{2}\left(\frac{a^{*2}}{V_y}\right) - \frac{k-1}{2}V_y \tag{6.14}$$

The momentum equation is written as

or,
$$(p_x - p_y)A = \dot{m}\ (V_y - V_x)$$

$$= \rho_x V_x (V_y - V_x) = \rho_y V_y (V_y - V_x)$$

or,
$$\frac{p_x}{\rho_x V_x} - \frac{p_y}{\rho_y V_y} = V_y - V_x$$

or, $$\frac{1}{V_x}\left(\frac{k\,p_x}{\rho_x}\right) - \frac{1}{V_y}\left(\frac{k\,p_y}{V_y}\right) = k\,(V_y - V_x)$$

But $$\frac{k\,p_x}{\rho_x} = a_x^2 \text{ and } \frac{k\,p_y}{\rho_y} = a_y^2$$

therefore,

$$\frac{a_x^2}{V_x} - \frac{a_y^2}{V_y} = k\,(V_y - V_x) \qquad (6.15)$$

Introduction of equations (6.13), (6.14) in (6.14) gives

$$\frac{k+1}{2}\left(\frac{a^{*2}}{V_x}\right) - \frac{k-1}{2}V_x - \frac{k+1}{2}\left(\frac{a^{*2}}{V_y}\right) - \frac{k-1}{2}V_y = k\,(V_y - V_x)$$

or, $$\frac{k+1}{2}a^{*2}\left(\frac{1}{V_x} - \frac{1}{V_y}\right) + \frac{k-1}{2}(V_y - V_x) = k\,(V_y - V_x)$$

or, $$\frac{k+1}{2}a^{*2}\left(\frac{V_y - V_x}{V_x V_y}\right) + \frac{k-1}{2}(V_y - V_x) = k\,(V_y - V_x)$$

or, $$\frac{k+1}{2}a^{*2} + \frac{k-1}{2}V_x V_y = k\,V_x V_y$$

This expression on simplification and rearrangement yields

$$V_x V_y = a^{*2} \qquad (6.16)$$

This is known as Prandtl-Mayer relation.

or, $$\frac{V_x}{a^*}\,\frac{V_y}{a^*} = 1$$

Now $a^* = a_x^* = a_y^*$

therefore,

$$\frac{V_x}{a_x^*}\,\frac{V_y}{a_y^*} = 1$$

or, $$M_x^* M_y^* = 1 \qquad (6.17)$$

This is also useful form of the Prandtl-Mayer relation

From the above equations (6.16) and (6.17), it is evident that if the flow at x is supersonic, the flow at y must be subsonic, and vice versa.

6.5 Working formulas for Normal shock

Generally the upstream Mach Number (M_x) in a given problem is known and it is desired to determine the downstream Mach Number (M_y) of the shock wave (See equation 6.12)

Equation (6.12) is most valuable in writing the basic equation in a form for easy tabulation.

Static pressure ratio across the shock is calculated by using the equation (6.9) and (6.12). The value of M_y^2 from equation (6.12) is substituted in equation (6.9)

$$\frac{p_y}{p_x} = \frac{1 + k\,M_x^2}{1 + \dfrac{k\,M_x^2 + \dfrac{2k}{k-1}}{\dfrac{2k}{k-1}M_x^2 - 1}}$$

or,
$$\frac{p_y}{p_x} = \frac{(1 + k\,M_x^2)\left[\left(\dfrac{2k}{k-1}\right)M_x^2 - 1\right]}{\dfrac{2k}{k-1}M_x^2 - 1 + \dfrac{2k}{k-1} + k\,M_x^2}$$

Simplifying and rearranging denominator.

$$\frac{p_y}{p_x} = \frac{(1 + k\,M_x^2)\left[\left(\dfrac{2k}{k-1}\right)M_x^2 - 1\right]}{\dfrac{k+1}{k-1} + \left(\dfrac{k+1}{k-1}\right)k\,M_x^2}$$

$$= \frac{(1 + k\,M_x^2)\left[\left(\dfrac{2k}{k-1}\right)M_x^2 - 1\right]}{\left(\dfrac{k+1}{k-1}\right)(1 + k\,M_x^2)}$$

$$= \frac{\dfrac{2k}{k-1}M_x^2 - 1}{\dfrac{k+1}{k-1}} = \frac{2kM_x^2}{k+1} - \frac{k-1}{k+1} \qquad (6.18)$$

Similarly taking equation (6.12) and (6.6) we will get the static temperature ratio across the shock

$$\frac{T_y}{T_x} = \frac{1 + \dfrac{k-1}{2} M_x^2}{1 + \left(\dfrac{k-1}{2}\right)\left[\dfrac{\dfrac{M_x^2 + \dfrac{2}{k-1}}{\dfrac{2k M_x^2}{k-1}}}{\dfrac{2k M_x^2}{k-1} - 1}\right]}$$

Simplifying only the denominator of the above equation

$$\frac{\dfrac{2\,k\,M_x^2}{k-1} - 1 + \dfrac{k-1}{2} M_x^2 + 1}{\dfrac{2k\,M_x^2}{k-1} - 1}$$

$$= \frac{M_x^2\,(k+1)^2}{2\,(k-1)} \Big/ \frac{2k\,M_x^2}{k-1} - 1$$

Thus,

$$\frac{T_y}{T_x} = \frac{\left(1 + \dfrac{k-1}{2} M_x^2\right)\left(\dfrac{2k}{k-1} M_x^2 - 1\right)}{\dfrac{(k+1)^2}{2\,(k-1)} M_x^2} \tag{6.19}$$

The stagnation pressure ratio p_{oy}/p_{ox} can be obtained from the following relation

$$\frac{p_{oy}}{p_{ox}} = \left(\frac{p_{oy}}{p_y}\right)\left(\frac{p_y}{p_x}\right)\left(\frac{p_x}{p_{ox}}\right)$$

We have

$$\frac{p_{oy}}{p_y} = \left(\frac{T_{oy}}{T_y}\right)^{k/k-1} = \left(1 + \frac{k-1}{2} M_y^2\right)^{k/k-1}$$

$$\frac{p_{ox}}{p_x} = \left(\frac{T_{ox}}{T_y}\right)^{k/k-1} = \left(1 + \frac{k-1}{2} M_x^2\right)^{k/k-1}$$

and $$\frac{p_y}{p_x} = \frac{2\,k\,M_x^2}{k+1} - \frac{k-1}{k+1}$$

Therefore,

$$\frac{p_{oy}}{p_{ox}} = \left[\frac{1+\frac{k-1}{2}M_y^2}{1+\frac{k-1}{2}M_x^2}\right]^{k/k-1} \left[\frac{2k\,M_x^2}{k+1} - \frac{k-1}{k+1}\right]$$

Substituting the value of M_y^2 from equation (6.12)

$$\frac{p_{oy}}{p_{ox}} = \left[\frac{1+\left(\frac{k+1}{2}\right)\left(M_x^2+\frac{2}{k-1}\right)}{\frac{2k}{k+1}M_x^2-1}\right]^{k/k-1} \left[\frac{\frac{2k\,M_x^2}{k+1}-\frac{k-1}{k+1}}{\left[1+\frac{k-1}{2}M_x^2\right]^{k/k-1}}\right]$$

Simplifying and algebraic manuplation, we get

$$\frac{p_{oy}}{p_{ox}} = \left[\frac{\frac{k+1}{2}M_x^2}{1+\frac{k-1}{2}M_x^2}\right]^{k/k-1} \left[\frac{2k}{k+1}M_x^2-\frac{k-1}{k+1}\right]^{1/k-1} \tag{6.20}$$

From the continuity equation

$$\frac{\rho_y}{\rho_x} = \frac{V_x}{V_y} = \left(\frac{p_y}{p_x}\right) \Big/ \left(T_y/T_x\right) = \left(\frac{p_y}{p_x}\right)\left(\frac{T_x}{T_y}\right)$$

Substituting the value of p_y/p_x and T_y/T_x from equation (6.18) and (6.19). we get

$$\frac{\rho_y}{\rho_x} = \frac{V_x}{V_y} = \frac{(k+1)\,M_x^2}{(k-1)\,M_x^2+2} \tag{6.21}$$

Finally, it is desirable to have an expression for p_{oy}/p_x. This, along with the pressure ratio already worked out, makes it possible, by simple combinations, to obtain any ratio of pressures or stagnation pressures.

$$\frac{p_{oy}}{p_x} = \left(\frac{p_{oy}}{p_y}\right)\left(\frac{p_y}{p_x}\right)$$

$$= \left[1 + \frac{k-1}{2} M_y^2\right]^{k/k-1} \left[\frac{2k\,M_x^2}{k+1} - \frac{k-1}{k+1}\right]$$

$$= \left[1 + \frac{k-1}{2}\left(\frac{M_x^2 + \frac{2}{k-1}}{\frac{2k}{k-1} M_x^2 - 1}\right)\right]^{k/k-1} \left(\frac{2k\,M_x^2}{k+1} - \frac{k-1}{k+1}\right)$$

$$= \left(\frac{k+1}{2} M_x^2\right)^{k/k-1} \left(\frac{2k\,M_x^2}{k+1} - \frac{k-1}{k+1}\right)^{1/1-k} \qquad (6.22)$$

6.6 The Rankine-Hugoniot relation

In the above analysis M_x was chosen as the independent variable. However it is often convenient to choose the pressure ratio (p_y/p_x) as the independent variable. The Rankine-Hugoniot relations express the density ratio and the temperature ratio on the two sides of a normal shock wave as a function of the pressure ratio.

Rewriting the equation (6.18)

$$\frac{2k}{k+1} M_x^2 = \frac{p_y}{p_x} + \frac{k-1}{k+1}$$

$$M_x^2 = \left(\frac{p_y}{p_x} + \frac{k-1}{k+1}\right)\left(\frac{k+1}{2k}\right) \qquad (6.23)$$

Subtituting the value of M_x^2 from equation (6.22) in equation (6.19) and simplifying we get

$$\frac{T_y}{T_x} = \frac{p_y/p_x\left[1 + \left(\frac{k-1}{k+1}\right)\frac{p_y}{p_x}\right]}{p_y/p_x + \frac{k-1}{k+1}} \qquad (6.24)$$

We know that from gas relation

$$\rho_y/\rho_x = \left(\frac{p_y}{p_x}\right)\left(T_y/T_x\right)$$

$$= \left(\frac{p_y}{p_x}\right)\left(\frac{T_x}{T_y}\right)$$

$$= \frac{\left(\frac{p_y}{p_x}\right)\left[\frac{p_y}{p_x} + \frac{k-1}{k+1}\right]}{\left(\frac{p_y}{p_x}\right)\left[1 + \frac{k-1}{k+1}\frac{p_y}{p_x}\right]} \qquad \text{[From equation 6.24]}$$

$$= \frac{\frac{p_y}{p_x} + \frac{k-1}{k+1}}{1 + \frac{k-1}{k+1}\frac{p_y}{p_x}}$$

Multiplying and dividing by $\frac{k+1}{k-1}$, we get

$$\frac{\rho_y}{\rho_x} = \frac{\frac{p_y}{p_x}\left(\frac{k+1}{k-1}\right)+1}{\frac{k+1}{k-1} + p_y/p_x}$$

Solving for p_y/p_x, we get

$$\frac{p_y}{p_x} = \frac{\rho_y/\rho_x\left(\frac{k+1}{K-1}\right)-1}{\left(\frac{k+1}{K-1}\right)-\rho_y/\rho_y} \tag{6.25}$$

The equation (6.25) is known as Rankline-Hugoniot relation and extensively used for analysis of shock strength.

This relation is plotted in Fig 6.4 together with the isentropic relation between pressure and density. It is seen that for diatomic gases the density can at most increase by a factor of six, whereas the pressure ratio may reach infinity. The curves indicate another important feature, namely that weak shocks are nearly isentropic; by "weak shock" we mean a shock in which the percentage rise in pressure is small.

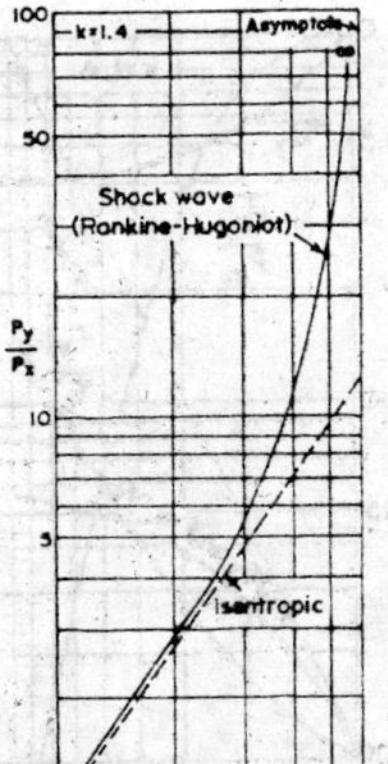

Fig. 6.4. The Rankine-Hugoniot curve for k=1.4

6.7 Working tables and charts

As in the case of isentropic flow, it is convenient to regard the Mach Number M_x as the forest parameter of the normal shock equations. Then, for any chosen value of M_x, the corresponding values of M_y, the ratios p_y/p_x, T_y/T_x, ρ_y/ρ_x, $\frac{p_{oy}}{p_{ox}}$ and V_x/V_y are easily tabulated from the above equations.

Use of the table for solving practical problems is preferable to using the equations directly, since in most cases this leads to lengthy trial and error procedure.

Important trends associated with the steady flow of an ideal gas across a normal shock wave can be determined by studying the Normal shock tables. These trends are summarized in table 6.1.

Fig 6.4 gives the important shock relations in graphical form for a gas with k = 1.4.

TABLE 6.1 Summary of normal wave characteristics

	Variable	Change across normal shock wave
1	Mach number, M	Decrease
2	Static Pressure, p	Increase
3	Stagnation on pressure p_o	Decrease
4	Static temperature, T	Increase
5	Stagnation temperature, T_o	Constant
6	Density	Increase
7	Velocity	Decrease

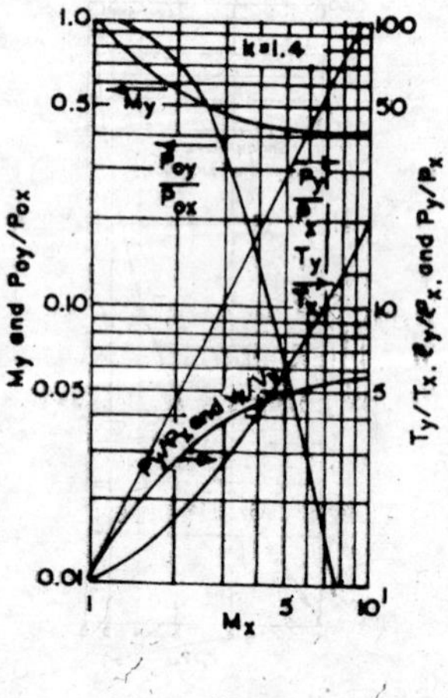

Fig. 6.5. Working curves for normal shock

6.8 IMPOSSIBILITY OF A SHOCK IN SUBSONIC FLOW

To find the entropy change across the shock, using the perfect gas formula

$$s_y - s_x = c_p \ln \frac{T_y}{T_x} - R \ln \frac{p_y}{p_x}$$

$$= c_p \ln \frac{T_y/T_x}{\left(p_y/p_x\right)^{k-1/k}}$$

$$= c_p \ln \frac{T_{oy}/T_{ox}}{\left(p_{oy}/p_{ox}\right)^{k-1/k}}$$

But $T_{oy} = T_{ox}$, so that

$$\frac{s_y - s_x}{R} = -\ln \frac{p_{oy}}{p_{ox}}$$

or, $$p_{oy}/p_{ox} = e^{-\Delta S/R} \qquad (6.26)$$

Substituting the value of p_{o_y}'/p_{o_x} from the equation (6.20)

$$\frac{s_y - s_x}{R} = \frac{k}{k-1} \ln \left[\frac{2}{K+1} M_x^2 + \frac{k-1}{k+1} \right]$$

$$+ \frac{1}{k-1} \ln \left[\frac{2k}{k+1} M_x^2 + \frac{k-1}{k+1} \right] \qquad (6.27)$$

Careful study of equation (6.27) indicates that for gases with $1 < k < 1.67$ the entropy change is always positive when M_x is greater than unity and is always negative when M_x is less than unity. The general form of equation (6.27) is shown graphically in Fig 6.6.

For a perfect gas only the show from supersonic to subsonic speed is possible. (Impossible subsonic to supersonic according to second law of thermodynamics).

Therefore shock waves cannot develop in a subsonic flow.

This fact is also corroborated by equation (6.27). For a decrease in entropy this equation gas

$$\frac{p_{oy}}{p_{ox}} > 1$$

Which implies that there is an increase in the stagnation pressure, but this is not possible in an adiabatic flow under consideration.

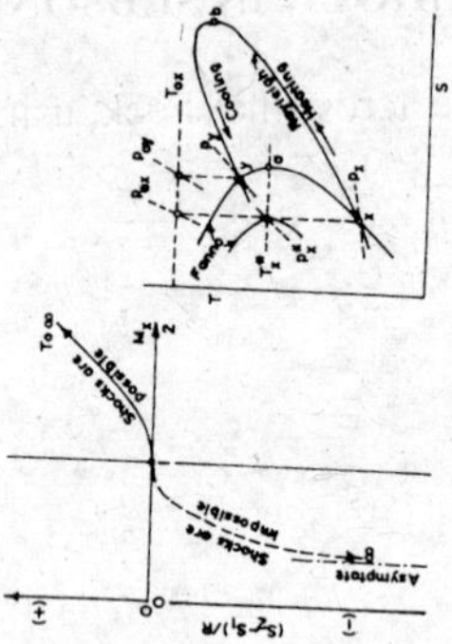

Fig. 6.6. Entropy change across normal shock wave

6.9 WEAK SHOCK WAVES

In order to discuss the possible simplification that may result from assuming that a shock wave has a very small pressure rise across it (that it is a "weak" shock wave), it is useful to introduce a parameter called shock strength.

6.9.1 Strength of shock wave

Strength of shock wave is defined as the ratio of pressure rise to initial pressure. It is denoted by P.

$$P = \frac{p_y - p_x}{p_x} = \frac{p_y}{p_x} - 1 \tag{6.28}$$

Substituting the value of $\frac{p_y}{p_x}$ from equation (6.18) in the above equation and rearranging yields

$$P = \frac{2k}{k+1}\left(M_x^2 - 1\right) \tag{6.29}$$

From the equation (6.29) it appears that the term $(M_x^2 - 1)$ is directly proportioal to the shock strength. We can say that the $(M_{x2} - 1)$ is an indicator of shock strength and shock wave depends upon Mx. Similarly the shock strength can be expressed in terms of density ratio through Rankline-Hygoniot relation (6.25). Thus, expressing py/px in terms of ρ_y/ρ_x. We get

$$p = \frac{p_y}{p_x} - 1 = \frac{\left(\frac{2k}{k-1}\right)\left(\frac{\rho_y}{\rho_x} - 1\right)}{\frac{2}{k-1} - \left(\frac{\rho_y}{\rho_x} - 1\right)} \tag{6.30}$$

For weak shocks i.e. for small values of the density ratio (ρ_y/ρ_x) following approximate relation is obtained

$$P \propto \left(\frac{\rho_y}{\rho_x} - 1 \right) \tag{6.31}$$

For air k = 1.4, the term of the equation $\left(\frac{2}{k-1} + 1 \right)$ in the denominator is equal to 6. It means that if ρ_y/ρ_x reaches a value of 6 the strength of the shock is infinity. Again if the process had been isentropic the pressure ratio for an optimum denisty ratio of 6 would be small, and shock strength would also be of the same order. This again stresses the point that weak shocks are nearly isentropic.

6.9.2 Shock of vanishing strength

Suppose that the shock strength approaches zero. Then from equation (6.29) the value of M_x approaches unity. Furthermore from equation (6.30), M_y approaches unity, and the density change across the shock approaches zero.

A shock of infinitestimal strength is therefore identical with a sound wave. Since M_x is unity for such a shock, it follows that a very weak pressure disturbance travels with the speed of sound with respect to the fluid in which it is propogating. An infinitestimal plane pressure wave, is, therefore, the limiting case of a normal shock. The shock itself travels with a speed greater than the speed of sound in the fluid into which it is propogating, the ratio of the two speeds being M_x.

Mathematically,

$$M_x \approx 1$$

$$\frac{\rho_y}{\rho_x} \approx 1$$

This condition leads to

$$p_y/p_x \approx 1, \frac{T_y}{T_x} \approx 1, \frac{p_{oy}}{p_{ox}} \approx \text{ and } \frac{\Delta S}{R} = 0$$

6.9.3 Irreversibility of weak shock

Entropy increase in terms of $\left(M_x^2 - 1 \right)$ by using the basic equation of perfect gas, density ratio from equation (6.21) and using the definition of weak shock strength, we get

$$\frac{s_y - s_x}{R} = \frac{2}{3} \frac{k}{(k+1)^2} \left(M_x^2 - 1 \right)^3$$

$$-\frac{2k^2}{(k+1)^2}\left(M_x^2-1\right)^4 + - - - \qquad (6.32)$$

Thus, for weak shocks, the entropy increase is of the third order with respect to the shock strength. As a first approximation, therefore, the irreversibility connected with weak shocks may be ignored, and the isentropic relation between pressure and density may be used for connecting the states on both sides of the shock.

This result explain why in Fig 6.4 the isentropic curve and Rankline-Hugoniot curve have the same slope and the same radius of curvature at the lower left hand corner of the chart. It explains also why in Fig 6.5 there is scarcely any loss in stagnation pressure across the shock for initial Mach numbers less than about 1.25.

6.10 SUPERSONIC PITOT TUBE

The pitot tube has long been an important instrument for measuring the velocities of subsonic into a supersonic flow stream, a shock wave is formed ahead of the tube. Because a shock wave gives rise to an increase in entropy, the isentropic flow relation derived earlier cannot be used. Thus a different relation becomes necessary and this is called the Rayleigh-Pitot equation.

6.10.1 Rayleigh supersonic pitot tube formula

Let us consider a shock wave ahead of a pitot tube as shown in Fig 6.7 (The shock wave can be seen with the help of photographs). In front of the tube the shock wave is normal and hence the air entering through the stagnation streamline will pass through a normal shock and will thus decelerate, becoming subsonic. The flow before the shock wave may be assumed to be isentropic. It then experiences an increase in entropy across the wave and is decelerated. After the shock wave, the flow is decelerated to the stagnation point p_{o_y}. Because the upstream and downstream conditions of a normal shock wave are defined by the two intersection points of Rayleigh and Fanno lines (Fig 6.8).

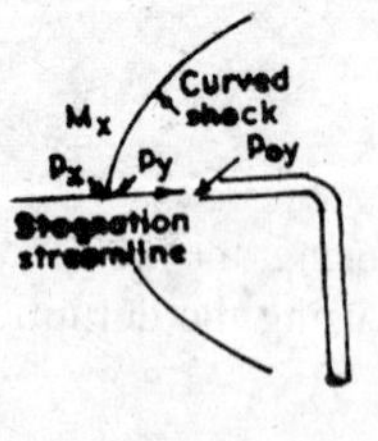

Fig. 6.7 Supersonic pitot tube

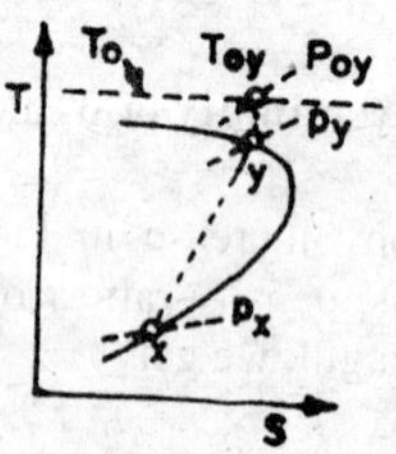

Fig. 6.8 T-S diagram

The p_{oy}/p_x can be wrtten as

$$\frac{p_{oy}}{p_x} = \frac{p_{oy}}{p_y}\frac{p_y}{p_x}.$$

Now, $\frac{p_{oy}}{p_y}$ is expressed in terms of M_x by equations (2.16) and (6.12) and p_y/p_x is given in terms of M_x by equation (6.18)

$$\frac{p_y}{p_x} = \frac{2k}{k+1}M_x^2 - \frac{k-1}{k+1}$$

Introducing these expressions, we get the Rayleigh pitot tube formula

$$\frac{p_{oy}}{p_x} = \left(\frac{k+1}{2}M_x^2\right)^{k/k-1}\left(\frac{2k}{k+1}M_x^2 - \frac{k-1}{k+1}\right)^{1/k-1} \tag{6.33}$$

Fig 6.9 shows the plot of equation (6.33)

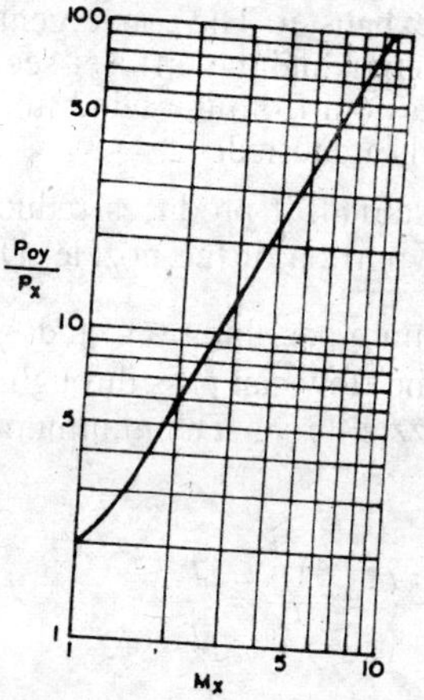

Fig. 6.9 Plot of Rayleigh supersonic pitot tube formula

6.11 ONE DIMENSIONAL SUPERSONIC DIFFUSERS

At first thought it might appear that a supersonic diffuser could be designed as though it were the reverse of a converging-diverging nozzle. Two difficulties arise, however. First, if the diffuser is in a closed system and is preceeded by a supersonic nozzle, frictional effects between the nozzle and the diffuser require that the diffuser throat be somewhat larger than the nozzle throat. If the diffuser throat is made slightly too small, supersonic flow will not be attained in the nozzle; and, if the diffuser throat is made slightly too large, there will necessarily be a shock somewhere within the diffuser. Indeed, even if the two throats did match perfectly, it appears that the combined system would be unstable and that flow oscillations would occur.

A second and more serious difficulty arises. Most flow systems start from rest and accelerate to the operating velocity. During the starting period normal shock waves pass through the system. Across a normal shock there is no change in flow rate or in stagnation temperature. The product $A_x^* \; P_{ox}$ must be equal to the product $A_y^* \; p_{oy}$. As the stagnation pressure is reduced by a normal shock, it follows that there is a corresponding increase in the minimum area through which the flow can be made to pass. The design problem raised by this aspect of the process will be discussed here.

6.11.1 Supersonic wind tunnel diffusers

A supersonic wind tunnel system of Fig 6.10 consists of a convergent divergent nozzle, a test section and a diffuser. The stagnation pressure and temperature, p_o and T_o are constant. The back pressure at the exit of the wind tunnel is controlled by an exhauster. The convergent divergent nozzle produces a supersonic speed in test section and the diffuser section reduces the velocity to a minimum so that it can pass out from the wind tunnel without noise. In the test section aerodynamic models are tested.

When the exhauster started, it produces reduced pressure at nozzle exit a shock moves in diverging portion of the nozzle. During this period, the nozzle throat is passing the maximum possible flow, and, since $A^* \; p_o$ is constant across the shock, it follows that the flow can pass through the diffuser throat only if the latter is larger than the nozzle throat. The minimum area of the diffuser throat is therefore given by

$$\frac{A_{minimum\ diffuser\ throat}}{A_{nozzle\ throat}} = \frac{A_y^*}{A_x^*} = \frac{p_{ox}}{p_{oy}} \tag{6.34}$$

Where p_{oy}/p_{ox} is the stagnation pressure ratio for a shock at the test section ratio for a shock at the test section Mach Number. The limiting contraction ratio A_x/A_y^* for the diffuser is accordingly given by

$$\frac{A_x}{A_y^*} = \frac{A_x}{A_x^*}\frac{A_x^*}{A_y^*} = \frac{A_x}{A_x^*}\frac{p_{oy}}{p_{ox}}$$

The area ratio A_x/A_x^* is given in terms of M_x by the isentropic formula and the ratio p_{oy}/p_{ox} is given in terms of M_x by the shock formula (equation 6.20)

Hence $\frac{A_x^*}{A_y^*}$ depends only on the test section Mach number. This area ratio is plotted versus M_x in Fig 6.11, together with a curve showing the area contraction which would be possible if the stream at Mx could be diffused without shock.

(a) Most favorable starting condition.

(b) Best operating condition.

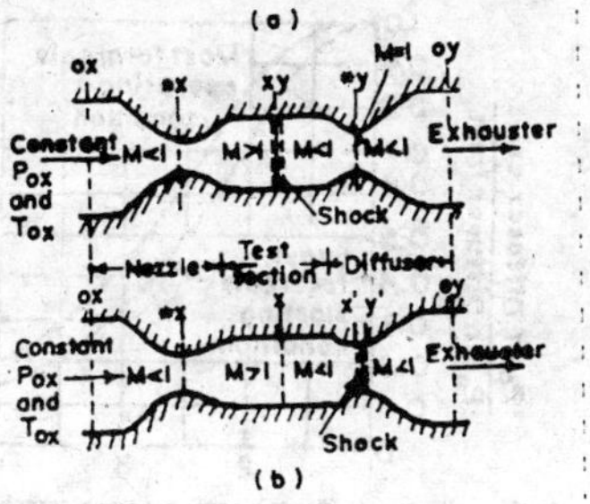

Fig. 6.10 Starting of supersonic wind tunnel diffuser

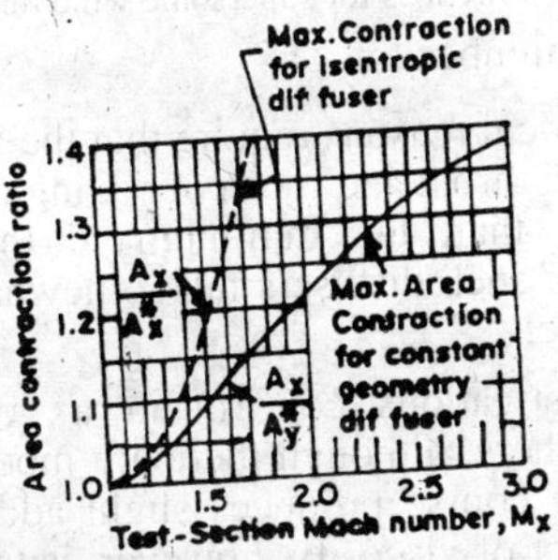

Fig. 6.11 Graph between Area contraction ratio v/s test section Mach Number M_x

Test section Mach number M_x

During the operating condition, shock to move through the test section and to be swallowed by the diffuser throat. For best efficiency, the system and exhauster should be matched so that the equilibrium position of the shock is in the diffuser throat, as the shock will then occur at the minimum possible Mach Number in the diffuser. The best operating condition is shown in Fig 6.10(b). The stagnation pressure loss for best operation is less than for starting up because of the lower Mach Number at which the shock occurs. Hence the pressure ratio required of the exhauster is determined by the starting condition rather than by the operating condition. A comparison of the two limiting stagnation pressure ratios is shown in Fig 6.12. The lower curve represents also the operating condition of a diffuser having no contraction whatsoever, i.e., a normal shock followed by isentropic subsonic diffusion. For low test section Mach Number it is evident that the normal shock is so efficient as hardly to justify the use of a converging diverging difffuser.

In practice the shock is maintained slightly down stream of the diffuser throat during operation. This is done because with a fixed exhaust pressure the shock is unstable in the converging portion of the diffuser. To obtain supersonic flow in the test section, it would be necessary to lowers th exhaust pressure to the minimum value required for starting, as determined from Fig 6.12.

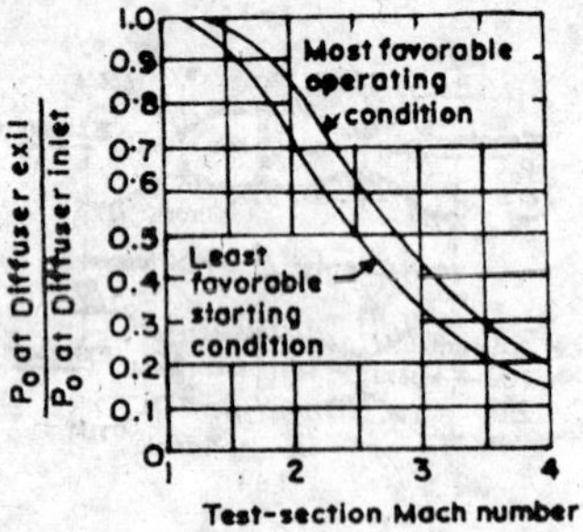

Fig. 6.12 Stagnation pressure loss for supersonic wind tunnel diffuser of fixed geometry.
Test section Mach number

Thus, because practical reasons require that the best design be compromised by an enlargement of the throat and by an operating condition with the shock at a Mach Number greater than the minimum in the passage, the practically attainable efficiencies of such diffusers fall somewhat short of the values which seen possible in principle.

Experimental investigations confirm in a general way the theoretical considerations outlined here, although there are modifications owing to viscous effects. Wall friction of course produces slight additional losses in stagnation pressure. The chief effect of viscosity, however, is to make the shock occur in a length of several passage diameters, rather than abruptly.

Some of the diffuser problems outlined here may be avoided by such means as

(a) Variable geometry diffusers

(b) Variable geometry diffusers in conjunction with variable geometry nozzles.

(c) driving the shock through the diffuser throat by means of a large amplitude pressure pulse, and

(d) taking advantage of effects which are not one dimensional.

6.11.2 Diffuser efficiency

Supersonic diffusers have applications in supersonic wind tunnels, compressors and supersonic inlets of aircraft engines and missiles.

The following assumptions are used for defining the diffuser efficiency

(a) The diffusion occurs in two stages

(i) through a normal shock at the entry section

(ii) through isentropic deceleration of the subsonic flow after the shock in the diverging passage.

(b) Perfect gas relations are used.

(c) The velocity at the diffuser exit is negligibly small, i.e. $p_{o_2} \approx p_2$

The reversible and irreverisible diffusion processes are shown by 1-2 s and 1 – 2 on the h – s diagram in Fig 6.13. Therefore,

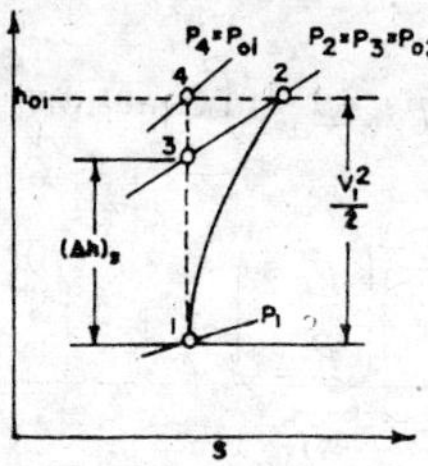

Fig. 6.13

$$\eta_D = \frac{(\Delta h)_S}{V_1^2/2} = \frac{h_3 - h_1}{h_2 - h_1}$$

Where State 1 is the actual state entering the diffuser, 2 is the actual state leaving the diffuser, and 3 is a fictitious state at the actual leaving pressure but at the entering entropy.

$$\eta_D = \frac{c_p (T_3 - T_1)}{V_1^2/2} = \frac{T_1 (T_3/T_1 - 1)}{V_1^2/2\, c_p}$$

$$= \frac{\left(\dfrac{p_2}{p_1}\right)^{(k-1)/k} - 1}{V_1/2\, c_p\, T_1} \qquad (6.35)$$

$$\eta_D = \frac{\left(p_2/p_1\right)^{(k-1)/k} - 1}{\dfrac{V_1^2}{\dfrac{2\, kRT}{k-1}}} = \frac{\left(p_2/p_1\right)^{(k-1)/k} - 1}{\left(\dfrac{k-1}{2}\right)\dfrac{V_1^2}{k\, RT}}$$

$$= \frac{\left(p_2/p_1\right)^{(k-1)/k} - 1}{\dfrac{k-1}{2} M_1^2} \qquad (6.36)$$

The efficiency is uniquely related to the stagnation pressure ratio for a given inlet Mach number. Therefore,

$$\frac{p_2}{p_1} = \frac{p_{o1}}{p_1}\cdot\frac{p_2}{p_{o1}} = \frac{p_{o1}}{p_1}\frac{p_{o2}}{p_{o1}} = \left(1+\frac{k-1}{2}M_1^2\right)^{k/(k-1)} \times \frac{p_{o2}}{p_{o1}}$$

Therefore, equation (6.36) may be written as

$$\eta_D = \frac{\left(1+\frac{k-1}{2}M_1^2\right)\left(\frac{p_{o2}}{p_{o1}}\right)^{(k-1)/k} - 1}{\frac{k-1}{2}M_1^2} \tag{6.37}$$

Substituting for $\frac{p_{o2}}{p_{o1}} = p_{oy}/p_{ox}$ in equation (6.37)

$$\eta_D = \frac{\left(1+\frac{k-1}{2}M_1^2\right)\left(\frac{p_{oy}}{p_{ox}}\right)^{(k-1)/k} - 1}{\frac{k-1}{2}M_1^2} \tag{6.38}$$

Substituting the value of p_{oy}/p_{ox} from equation (6.20) into equation (6.38) and rearranging, we get

$$\eta_D = \frac{\left(\frac{k+1}{k-1}\right)}{\left(\frac{2k}{k+1}M_1^2 - \frac{k-1}{k+1}\right)^{1/k}} - \left(\frac{2}{k-1}\right)\frac{1}{M_1^2} \tag{6.39}$$

Equations (6.38) and (6.39) show that the diffuser efficiency is only a function of the entry Mach number.

Corresponding to the two curves of Fig (6.12), these are plotted in Fig (6.14), with the help of above equation, the maximum efficiencies for fixed geometry diffusers and normal shock diffusers.

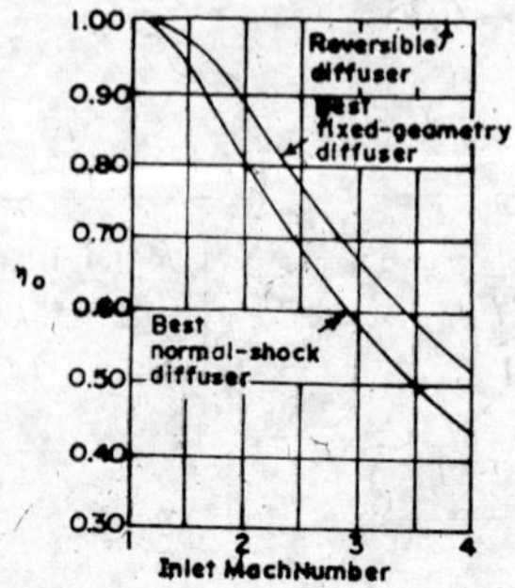

Fig. 6.14 Maximum efficiency of one dimensional supersonic diffusers.

ILLUSTRATIVE EXAMPLES

6.1 Air flowing through a nozzle encounters a shock. The Mach number upstream of the shock is 1.8 and the static pressure downstream of the shock is 425°C. How much has the velocity changed across the shock ? Assume k = 1.4.

Solution From normal shock tables

At $M_x = 1.8, \frac{\rho_x}{\rho_y} = 0.42837 = \frac{V_y}{V_x}, \frac{T_y}{T_x} = 1.5316$

$$\frac{T_x}{T_y} = 0.6529$$

$$T_x = 0.6529 \times 698 = 455.7\ K.$$

$$a_x = 20.1\sqrt{455.7} = 429.1\ m/s$$

$$V_x = M_x a_x = 1.8 \times 429.1 = 772.4\ m/s$$

$$V_y = 0.42387\ V_x = 0.42387 \times 772.4 = 327.4\ m/s$$ Ans

$$\therefore \quad V_x - V_y = 772.4 - 327.4 = 445\ m/s$$ Ans

6.2 A stream of air with a Mach number of 2, pressure of 0.7 bar and temperature 15°C passes through a normal shock. Calculate the final velocity and stagnation pressure.

Solution

Given $M_x = 2,\ p_x = 0.7\ bar,\ T_x = 288\ K$

From isentropic tables,

At $M_x = 2, \frac{p_x}{p_{ox}} = 0.1278$

$$\therefore \quad p_{ox} = \frac{0.7}{0.1278} = 5.477\ bar$$

$$a_x = 20.05\sqrt{288} = 340.25\ m/s$$

$$V_x = M_x a_x$$

$$= 2 \times 340.25 = 680.519\ m/s$$

From normal shock tables

At $M_x = 2$

$$\frac{V_x}{V_y} = 2.667, \frac{p_{oy}}{p_{ox}} = 0.7209$$

$$V_y = \frac{680.519}{2.667} = 255.16\ m/s$$

$$p_{oy} = 5.477 \times 0.7209 = 3.948\ bar$$

Ans

6.3 A normal shock wave is standing in the test section of supersonic wind tunnel. Upstream of the wave, $M_x = 3$, $p_x = 0.5$ bar, and $T_x = 200$ K.

Find M_y, p_y, T_y and V_y downstream of the wave.

Solution From normal shock tables

At $\quad M_x = 3, \frac{p_y}{p_x} = 10.33, \frac{T_y}{T_x} = 2.679$

and $\quad M_y = 0.4752$

$$\therefore \quad p_y = p_x \times 10.33 = 0.5 \times 10.33 = 5.165\ bar$$

$$T_y = 2.679 \times 200 = 535.8\ K$$

$$a_y = \sqrt{k\,RT_y} = \sqrt{1.4 \times 287 \times 535.8} = 463.98\ m/s$$

$$a_x = \sqrt{k\,RT_x} = \sqrt{1.4 \times 287 \times 200} = 284\ m/s$$

$$V_y = M_y\,a_y = 0.4752 \times 463.98 = 220.486\ m/s$$

Ans

6.4 A converging diverging nozzle has an exit ratio of 3. Air enters the nozzle with a stagnation pressure of 6.5 bar and a stagnation temperature of 360 K. Assuming a normal shock at the exit plane and considerding the flow through the diverging portion of the nozzle as supersonic, determine the Mach number, pressure and temperature.

Solution From isentropic tables

At $\quad \frac{A_x}{A^*} = 3,\ M_x = 2.65$

$$\frac{p_x}{p_{ox}} = 0.0463, \quad \frac{T_x}{T_{ox}} = 0.4158$$

$$p_x = 6.5 \times 0.0463 = 0.30095\ bar$$

$$T_x = 360 \times 0.4158 = 149.68\ K$$

From shock tables

At $\quad M_x = 2.65,\ M_y = 0.5008$

$$\frac{p_y}{p_x} = 8.0262, \quad \frac{T_y}{T_x} = 2.2901$$

Therefore, $p_y = 8.0262 \times 0.30095 = 2.41548\ bar$

$T_y = 2.2901 \times 149.68 = 342.78\ K$ Ans

6.5 Consider the subsonic-supersonic flow through a convergent divergent nozzle. The reservoir pressure and temperature are 10 bar and 300 K; respectively. There are two locations in the nozzle where $\frac{A}{A^*} = 6$, one in the convergent section and the other in the divergent section. At each location, calculate M, p, T and V.

Solution In the convergent section, the flow is subsonic.

For subsonic flow with $\frac{A}{A^*} = 6$, M = 0.097

$$\frac{p_o}{p} = 1.006, \quad \frac{T_o}{T} = 1.002. \textit{ Hence}$$

$$p = \frac{p}{p_o} \times p_o - \frac{1}{1.006} \times 10 = 9.94\ bar$$

$$T = \frac{T}{T_o} \times T_o = \frac{1}{1.002} \times 300 = 299.4\ K$$

$$a = \sqrt{kRT} = \sqrt{1.4 \times 287 \times 299.4} = 346.8\ m/s$$

$$V = M\,a = 0.097 \times 346.8 = 33.6\ m/s$$

In the divergent section, the flow is supersonic.

$$\frac{A}{A^*} = 6, \ M = 3.368, \quad \frac{p_o}{p} = 63.13, \quad \frac{T_o}{T} = 3.269$$

$$p = \frac{p}{p_o} \times p = \frac{1}{63.13} \times 10 = 0.1584\ atm.$$

$$T = \frac{T}{T_o} \times T_o = \frac{1}{3.269} \times 300 = 91.77\ K$$

$$a = \sqrt{kRT} = \sqrt{1.4 \times 287 \times 91.77} = 192\ m/s$$

$$V = M\,a = 3.368 \times 192 = 646.7\ m/s$$ Ans

6.6 Air from a reservoir at a pressure p_o flows through a converging diverging nozzle of throat area A^* and an exit area 2.54 A^*. A normal shock occurs beyond the throat where A = 1.45 A^*

(a) What is the exit Mach number ?

(b) What is the receiver pressure in terms of p_o ?

(c) What receiver pressure would produce isentropic supersonic flow without shocks ?

Solution

(a) $\frac{A}{A^*} = 1.45$ (Given)

Using isentropic tables

$M_x = 1.81$

Using shock tables. $M_y = 0.614$

and $p_{oy}/p_{ox} = 0.8082$

For isentropic flow downstream of the shock

$$\frac{A_y}{A_y^*} = 1.171$$

$$\text{Thus,} \quad \frac{A_e}{A_y^*} = \frac{A_y}{A_y^*} \times \frac{A_e}{A_y} = 1.171 \times \frac{2.5A_x^*}{1.45A_x^*}$$

Therefore, $M_e = 0.303$ and $\frac{p_e}{p_{oy}} = 0.9382$ Ans

(b) $\frac{p_e}{p_{ox}} = \frac{p_e}{p_{oy}} \times \frac{p_{oy}}{p_{ox}} = 0.9382 \times 0.8082 = 0.758$

(c) For $\frac{A}{A^*} = 2.5$,

Using isentropic tables

$M_e = 2.443$

and $\frac{p_e}{p_o} = 0.0640$ Ans

6.7 (a) Prove that for a normal shock

$$V_x V_y = a^{*2}$$

and hence sketch a curve between M_y and M_x.

Is an expansion or rare faction shock possible in perfect gas ? Why ?

(b) The following data refers to a supersonic wind tunnel.

Working fluid (R = 287 J/kg k, k = 1.4)

Nozzle throat area = 200 cm^2

Test section cross section = 337.5 cm^2

Determine the test section Mach number and diffuser throat area if a normal shock is located in the test section.

Solution

(a) See text.

(b) $\frac{A}{A^*} = \frac{337.5}{200} = 1.6875$

From isentropic tables

At $\frac{A}{A^*} = 1.6875, M_x = 2$ Ans

From normal shock tables.

At $M_x = 2, \frac{p_{oy}}{p_{ox}} = 0.721, M_y = 0.577.$

For normal shock in the test section

$$\frac{\dot{m}\sqrt{T_{ox}}}{A_x^* p_{ox}} = \frac{\dot{m}\sqrt{T_{oy}}}{A_y^* p_{oy}}$$

But for a normal shock $T_{ox} = T_{oy}$

and $\dot{m}$ = *constant*

Therefore, $A_x^* p_{ox} = A_y^* p_{oy}$

$$A_y^* = \left(\frac{p_{ox}}{p_{oy}}\right) A_x^* = \frac{200}{0.721} = 277.32\ cm^2$$ Ans

6.8 Air flows through a converging diverging nozzle with area ratio, $A_e/A_t = 3.5$. The upstream stagnation conditions are atmospheric; the back pressure is maintained by a vacuum pump. Determine the back pressure required to cause a normal shock to stand in the nozzle exit plane, and the flow speed leaving the shock.

Solution

$A_1/A_t = 3.5, T_o = 288\ K, p_o = 101\ kPa$

From isentropic tables

At $A_1/A_t = 3.5,\ M = 2.8, \frac{T_1}{T_{o1}} = 0.3894$

and $p_1/p_{o1} = 0.03685$

Therefore, $T_1 = 0.3894 \times 288 = 112.147\ k = T_x$

$$p_1 = 0.03685 \times 101 = 3.72\ kPa = p_x$$

From normal shock tables

At $\quad M_x = 2.8, M_y = 0..488, \dfrac{T_y}{T_x} = 2.451$

and $\quad p_y/p_x = 8.98$

Therefore, $T_y = 2.451 \times T_x = 2.451 \times 112.147 = 274.87\ K$

$$p_y = 8.98 \times p_x = 8.98 \times 3.72 = 33.4\ kPa \qquad \text{Ans}$$

$$a_y = \sqrt{k\,RT_y} = \sqrt{1.4 \times 287 \times 274.87} = 332.33\ m/s$$

$$V_y = M_y\,a_y = 0.488 \times 332.33 = 162.176\ m/s \qquad \text{Ans}$$

6.9 A fixed convergent divergent engine intake is designed so that at a maximum air flow a normal shock stand just upstream of the mouth in flight at a Mach number of 1.5. What is the ratio of mouth area to throat area ? Neglecting friction, find maximum possible mass flow rate into engine for 0.1 m^2 throat area, in flight at Mach number of 1.5 and at an altitude of 10000 m just before the shock is swallowed and at a slightly higher speed with shock swallowed.

Solution

(a)When the shock is detached, we have from shock tables. at $M_x = 1.5$, $M_y = 0.707$.

The flow then acceleratrs to unity. Therefore from isentropic tables at M = 0.707, $\dfrac{A}{A^*} = 1.094$. Ans

(b) When the shock is swallowed, i.e. it occurs at minimum cross section (i.e. at the throat of the passage).

The standard value of pressure and temperature at 10000 m are 26.4 kPa and 222.9 K respectively.

Therefore from isentropic tables.

At $\quad M = 1.5, \dfrac{p}{p_o} = 0.272, \dfrac{T}{T_o} = 0.689$

Therefore, $p_o = \dfrac{26.4}{0.272} = 97.058\ kPa$

and $\quad T_o = \dfrac{222.9}{0.689} = 323.51\ kPa$

We know from Fliegner's formula.

$$\frac{\dot{m}_{max}\sqrt{T_o}}{p_o A^*} = 0.040418$$

or, $$\frac{\dot{m}_{max}}{A^*} = \frac{0.040418 \times 97.058 \times 1000}{\sqrt{323.51}} = 218.1$$

$$\dot{m}_{max} = A^* \times 218.1 = 0.1 \times 218.1 = 21.81\ kg/s$$ Ans

6.10 Air is expanded in a nozzle from 20 bar, 200°C to 2 bar. The mass rate of flow through the nozzle is 5 kg/s. Assume the flow to be reversible and adiabatic. Determine

(a) What back pressure will cause a normal shock to stand in the exit plane of the nozzle. What is the mass flow rate under these conditions ?

(b) At what Mach number will the normal shock occur in the nozzle if the back pressure is 1.4 MPa ? Take $A_e = 2596\ mm^2$ and $A^* = 1345\ mm^2$.

(c) What back pressure will be required to cause subsonic flow throughout the entire nozzle with M = 1 at the throat.

(d) Determine the mass flow rate for a back pressure of 1.9 MPa.

Solution

(a) $$\frac{p_e}{p_o} = \frac{0.2}{2} = 0.1$$

From isentropic tables

At $p_e/p_o = 0.1, M_e = M_x = 2.16$

From shock tables

At $M_x = 2.16, M_y = 0.553, \frac{p_y}{p_x} = 5.276$

Therefore, $p_y = 5.276\,p_x = 5.276 \times 2 = 10.552\ bar$

Therefore the mass flow rate is same $\dot{m} = 5\ kg/s$ Ans

(b) Assuming $M_x = 1.85$; therefore $M_y = 0.6061$

$$\frac{p_{oy}}{p_{ox}} = 0.79, \frac{A_e}{A_x^*} = \frac{2596}{1345} = 1.9301$$

At $M_x = 1.85, A_x/A_x^* = 1.4971$

At $M_y = 0.6061, A_x/A_x^* = 1.1885$

Therefore, $\dfrac{A_e}{A_y^*} = \dfrac{A_e}{A_x^*} \times \dfrac{A_x^*}{A_y^*}$

$$= \frac{A_e}{A_x^*} \times \frac{A_x}{A_y^*} \times \frac{A_x^*}{A_x}$$

$$= 1.930 \times 1.1885 \times \frac{1}{1.4971} = 1.532$$

Therefore, at $A_e/A_y^* = 1.532$, $M_e \approx 0.42$

$$\frac{p_e}{p_{oy}} = 0.886$$

Therefore, $\dfrac{p_e}{p_{ox}} = 0.886 \times 0.79 = 0.6994$

$$p_e = 0.6994 \times 20 = 14 \; bar = 1.4 \; MPa$$

Hence the assumed Mach number is okay.

(c) $A_e/A^* = 1.93$, from table, $M_e = 0.3236$

and $\dfrac{p_e}{p_o} = 0.92912$

Therefore, $p_e = 0.92912 \times 20 = 18.58 \; bar$

(d) $p_e/p_{ox} = \dfrac{1.9}{2} = 0.95$, From table, $M_e = 0.268$

$$T_e = 0.9854 \times 473 = 466.1 \; K$$

$$a_e = \sqrt{k\,RT_e} = \sqrt{1.4 \times 287 \times 466.1} = 432.75 \; m/s$$

$$V_e = M_e\, a_e = 0.268 \times 432.75 = 115.977 \; m/s$$

$$\rho_e = p/RT_e = 20 \times 100/0.287 \times 466.1 = 14.95 \; kg/m^3$$

$$\dot{m} = \rho_e A_e V_e$$

$$= 14.95 \times 259 \times 10^{-6} \times 115.977 = 4.5 \; kg/s$$ Ans

6.11 Air flows through a converging diverging nozzle with area ratio, $A_e/A_t = 3.5$. The upstream stagnation conditions are atmoshperic; the back pressure is maintained by a vacuum system. Determine the range of back pressure for which a normal shock will occur within the nozzle, and corresponding mass flow rate is $A_t = 500\ mm^2$.

Solution

$$A_e/A_t = 3.5, T_o = 288\ K, p_o = 101\ kPa$$

A normal shock will occur within the nozzle for back pressure conditions in Regime II

For isentropic flow with $A_e/A_t = 3.5$.

$$M = 0.169, p/p_o = 0.9858 \therefore p_b = 96.6\ kPa$$

and $M = 2.8, p/p_o = 0.03695$

From normal shock tables

At $M_x = 2.8, M_y = 0.466, p_y/p_x = 8.980$

Thus $p_b = p_y = p_o \dfrac{p_x}{p_o} \times \dfrac{p_y}{p_x}$

$$= 101 \times 0.03685 \times 8.980 = 33.42\ kPa$$

33.4 kPa < p_b < 99.6 kPa (for normal shock in nozzle) Ans

Flow is choked throughout this regime. Thus

$$\dot{m} = \rho_t V_t a_t, \text{ therefore, At } M = 1$$

$$\frac{p_t}{p_o} = 0.5283$$

and $T_t/T_o = 0.8333$

$$p_t = 0.5283 \times 101 = 53.3583\ kPa$$

$$T_t = 0.8333 \times 288 \approx 240\ K$$

Therefore, $\rho_t = \dfrac{p_t}{RT_t} = \dfrac{53.3583}{0.287 \times 240} = 0.77468\ kg/m^3$

$$V_t = M a_t = 1 \times \sqrt{k RT_t} = \sqrt{1.4 \times 287 \times 240} = 310.535\ m/s$$

$$\dot{m} = \rho_t A_t V_t = 0.77468 \times 500 \times 10^{-6} \times 310.535$$

$$= 0.12028\ kg/s$$

Ans

6.12 A converging diverging nozzle with area ratio $A_e/A_t = 4$, is designed to expand air isentropically to atmospheric pressure. Determine the exit Mach number at design conditions, and the required value of inlet stagnation pressure. Determine the back pressures that correspond to the boundaries of Regimes I, II, III and IV. Sketch the plot of pressure ratio versus axial distance for this nozzle.

Solution

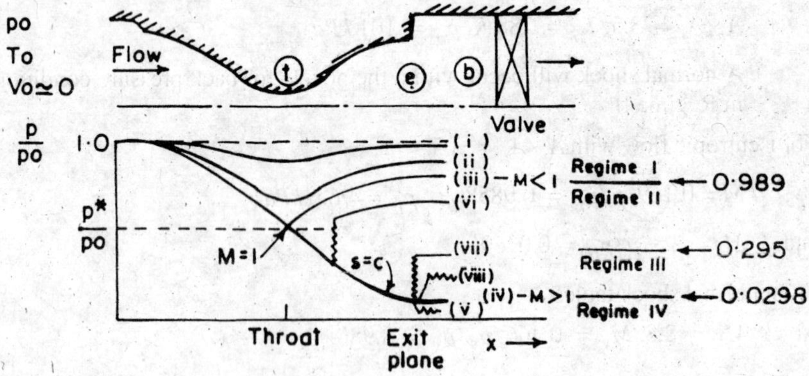

Fig. 6.15 Pressure distribution for flow in a converging-diverging nozzle as a function of back pressure. P. 6.12

From isentropic tables

At $\quad \dfrac{A_e}{A_t} = \dfrac{A_e}{A^*} = 4$

$$M_e = 0.147,\ p_e/p_o = 0.9887$$

and $\quad M_e = 2.94,\ p_e/p_o = 0.0298$

Then

$$M_{e\ design} = 2.94$$

$$p_{o\ design} = \frac{p_e}{0.0298} = \frac{101}{0.0298}$$

$$= 3389.26\ kPa \approx 3.39\ MPa$$ Ans

$$p_{b1} = \frac{p_e}{p_o} \times p_{od} = 0.9887 \times 3.39 = 3.35\ MPa$$ Ans

$$p_{b3} = 101\ kPa$$ Ans

For $\quad M_e = 2.94$, Using normal shock tables,

$$p_y/p_x = 9.917$$

Therefore, $p_{b2} = \dfrac{p_y}{p_x} \times p_{b3} = 9.917 \times 101 = 1\ MPa$ Ans

6.13 A converging-diverging nozzle has an exit area to throat area ratio of 2. Air enters this nozzle with a stagnation pressure of 1000 kPa and a stagnation tempr of 360 K. The throat area is 500 mm^2. The divergent section of the nozzle acts as a supersonic nozzle. Assume that a normal shock stands in the exit plane of the nozzle. Determine the static pressure, and temperature and the stagnation pressure just downstream of the normal shoke.

Assume that these is a normal shock wave standing at the point where M = 1.5. Determine the exit plane pressure, temperature, and Mach number. Assume isentropic flow except for the normal shock.

Solution

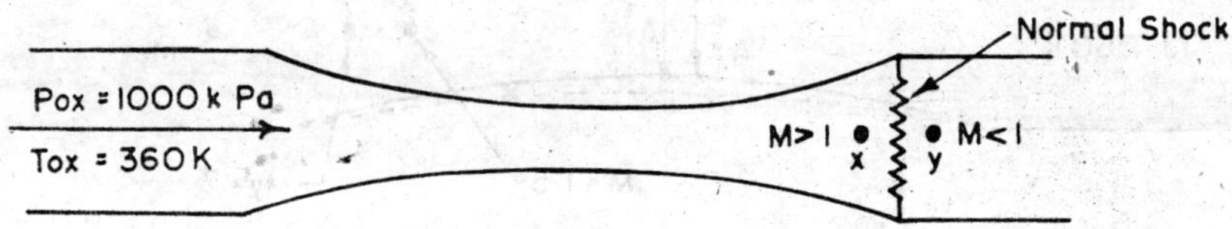

Fig. 6.16 Converging diverging nozzle P. 6.13

$$\frac{A}{A^*} = 2$$

Diverging section acting as a supersonic nozzle

$M_x = 2.197$

$p_x = 93.9$ kPa

$T_x = 183.2$ K

$p_{ox} = 1000\ kPa$

From isentropic nozzle

At $\frac{A_e}{A^*} = 2, M_e = M_x = 2.197, \frac{p_x}{p_{ox}} = 0.0939, \frac{T_x}{T_{ox}} = 0.5089$

$\therefore \quad p_x = 0.0939 \times 1000 = 93.9\ kPa = p_x$

$T_x = 0.5089 \times 360 = 183.2\ kPa = T_x$

From shock table.

At $M_x = 2.197,\ M_y = 0.547,\ \frac{p_y}{p_x} = 5.46$

$\frac{T_y}{T_x} = 1.854, \frac{p_{oy}}{p_{ox}} = 0.630$

Therefore,

$$p_y = 5.46 \times 93.9 = 512.7\ kPa$$

$$T_y = 1.854 \times 183.2 = 339.7\ K$$

$$p_{oy} = 0.630 \times 1000 = 630\ kPa$$

Assume that there is a normal shock wave standing at the point where M = 1.5. Determine the exit plane pressure, temperature, and Mach number. Assume isentropic flow except for the normal shock.

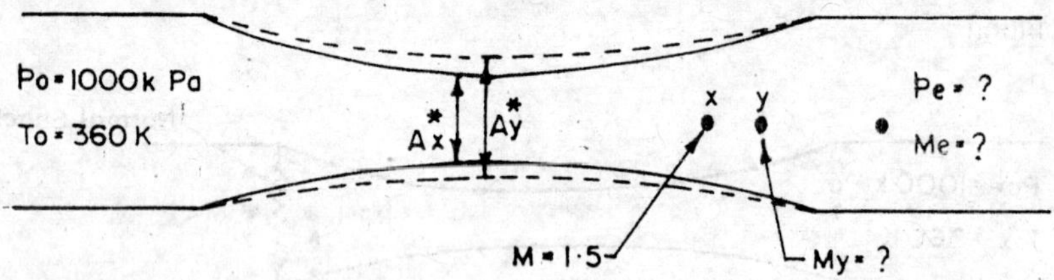

Fig. 6.17 Converging-diverging nozzle P. 6.13.

The flow is isentropic upto x

At $M_x = 1.5, \dfrac{p_x}{p_{ox}} = 0.2724, \dfrac{T_x}{T_{ox}} = 0.6897$

$$\frac{A_x}{A_x^*} = 1.1792$$

$\therefore \quad p_x = 0.2724\ (1000) = 272.4\ kPa$

$$T_x = 0.6897 \times 360 = 248.3\ K$$

From shock tables

At $M_x = 1.5, M_y = 0.7011, \dfrac{p_y}{p_x} = 2.4583,$

$$\frac{T_y}{T_x} = 1.32, \frac{p_{oy}}{p_{ox}} = 0.9290$$

$$p_y = 2.4583 \times 272.4 = 669.6\ kPa$$

$$T_y = 1.32 \times T_x = 1.32 \times 248.3 = 327.8\ K$$

$$p_{ox} = 0.9298\ p_{ox} = 0.9298 \times 1000 = 929.8\ kPa$$

$$T_{ox} = T_{oy} = 360\ K$$

From y to e the diverging section acts as a subsonic diffuser.

In solving this problem it is convenient to think of the flow at y as having come from an isentropic nozzle having a throat area A_y^*. Such a hypothetical nozzle is shown by the dotted line.

From the table of isentropic flow functions.

$$M_y = 0.7011, \frac{A_y}{A_y^*} = 1.0938, \frac{p_y}{p_{oy}} = 0.7202$$

$$\frac{T_y}{T_{oy}} = 0.9105$$

$$\frac{A_e}{A_x^*} = 2 \text{ also the flow from y to e isentropic}$$

$$\frac{A_e}{A_e^*} = \frac{A_e}{A_y^*} = \frac{A_e}{A_x^*} \times \frac{A_x^*}{A_x} \times \frac{A_x}{A_y} \times \frac{A_y}{A_y^*}$$

$$\frac{A_e}{A_y^*} = 2 \times \frac{1}{1.1762} \times 1 \times 1.0938 = 1.860$$

From the table of isentropic flow function for

$$\frac{A}{A^*} = 1.860, \text{ and } M < 1$$

$$M_e = 0.339, \frac{p_e}{p_{oe}} = 0.9222, \frac{T_e}{T_{oe}} = 0.9771$$

$$\frac{p_e}{p_{oe}} = \frac{p_e}{p_{oy}} = 0.9222$$

$$p_e = 0.9222 \times 929.8 = 857.5\ kPa$$

$$T_e = 0.9771 \times 360 = 351.7\ K$$

6.14 A convergent-divergent nozzle operates at off design condition while conducting air from a high pressure tank to a large container. A normal shock occurs in the divergent part of the nozzle at a section where the cross sectional area is 18.75 cm^2. If the stagnation pressure and stagnation temperature at the inlet of the nozzle are 0.21 MPa and 36°C respectively, and the throat area is 12.5 cm^2 and the exit area is 25 cm^2 estimate the exit Mach number exit pressure, loss in stagnation pressure, and entropy increase, during the flow between the tanks.

Solution

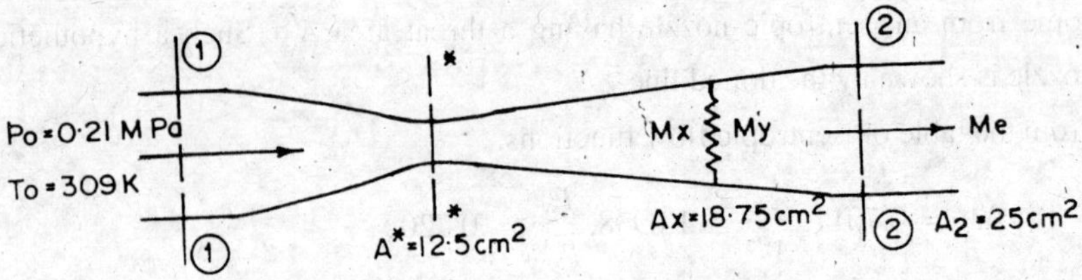

Fig. 6.18 Convergent-divergent nozzle P. 6.14

$$\frac{A_x}{A^*} = \frac{18.75}{12.5} = 1.5$$

Flow upto the shock is isentropic ∴ Isentropic table is used

$$\therefore \quad \frac{A}{A^*} = 1.5, M_x = 1.86, \frac{p_x}{p_{ox}} = 0.159$$

$$p_x = 0.159 \times 0.21 \times 10^3 = 33.4 \; kPa$$

From shock tables

$$M_x = 1.86, M_y = 0.604, \frac{p_y}{p_x} = 3.87$$

$$\frac{p_{oy}}{p_x} = 4.95, \frac{p_{oy}}{p_{ox}} = 0.786$$

$$p_y = 3.87 \times 33.4 = 129.3 \; kPa$$

$$p_{oy} = 4.95 \times 33.4 = 165.3 \; kPa$$

From the shock section to the exit of the nozzle, the flow is again isentropic

When $M_y = 0.604$ From the isentropic flow tables

$$\frac{A_y}{A^*} = 1.183$$

$$\therefore \quad \frac{A_2}{A^*} = \frac{A_2}{A_y} \times \frac{A_y}{A^*} = \frac{25}{18.75} \times 1.183 = 1.582$$

When $\frac{A_2}{A^*} = 1.582$, from the isentropic flow table

$$M_2 = 0.402$$ Ans

$$\frac{p_2}{p_{oy}} = 0.895 \qquad \therefore p_2 = 0.895 \times 165.3 = 147.94 \; kPa$$

Loss in stagnation pressure occurs only acroos the shock.

$$p_{ox} - p_{oy} = 210 - 165.3 = 44.7 \; kPa$$

Entropy increase, $s_y - s_x = +R \ln \dfrac{p_{ox}}{p_{oy}}$

$$= 0.287 \ln \frac{210}{165.3}$$

$$= 0.287 \times 0.239 = 0.0686 \; kJ/kg\, K$$ Ans

6.15 In a converging-diverging nozzle of exit to throat area ratio 2.5. Air at 2.5 bar & 117°C flows through a nozzle with a velocity of 250 m/s. If a normal shock stands at a point where the Mach number is 2; find the fluid properties at the exit plane.

Solution

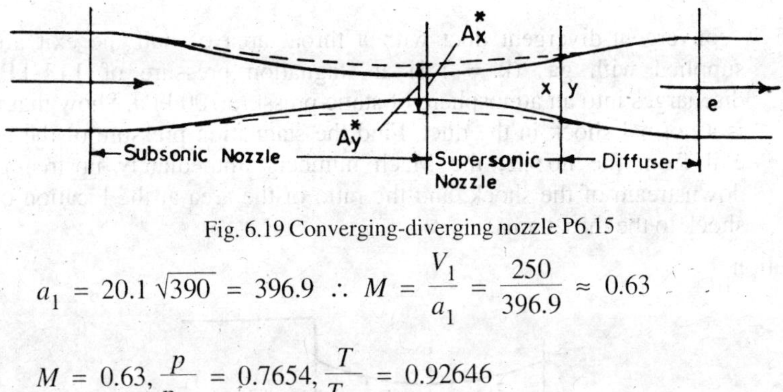

Fig. 6.19 Converging-diverging nozzle P6.15

$$a_1 = 20.1\sqrt{390} = 396.9 \;\therefore\; M = \frac{V_1}{a_1} = \frac{250}{396.9} \approx 0.63$$

At $\quad M = 0.63, \dfrac{p}{p_{ox}} = 0.7654, \dfrac{T}{T_{ox}} = 0.92646$

$$\therefore \quad p_{ox} = 2.5/0.7654 = 3.266 \; bar,$$

$$T_{ox} = \frac{390}{0.92646} = 421 \approx 420.96 \; K$$

$$\rho_{ox} = \frac{p_{ox}}{RT_{ox}} = \frac{3.266 \times 10^5}{287 \times 420.96} = 2.65 \; kg/m^3$$

At normal shock $M_x = 2 \;\therefore\; M_y = 0.57735$,

From shock table

$$\frac{p_y}{p_x} = 4.5, \frac{T_y}{T_x} = 1.6875$$

$$\frac{A_e}{A_e^*} = \frac{A_e}{A_y^*} = \frac{A_e}{A_x^*} \times \frac{A_x^*}{A_y^*} = 2.5 \times 0.72088 = 1.8022$$

$$\frac{p_{oy}}{p_{ox}} = 0.72088 = \frac{A_x^*}{A_y^*} \qquad \therefore p_{oy} = 3.266 \times 0.72088 = 2.354\ bar$$

From isentropic table for $\frac{A}{A^*} = 1.8022$ in subsonic range

$$M = 0.345, \frac{p_e}{p_{oy}} = 0.9209, \frac{T_e}{T_{oy}} = 0.97674, \frac{\rho_e}{\rho_{oy}} = 0.94297$$

$$\therefore \quad p_e = 0.9209 \times 2.354 = 2.168\ bar$$

$$T_e = 420.46 \times 0.97674 = 411.2\ K$$

$$\rho_e = 1.8\ kg/m^3$$

Ans

6.16 A convergent-divergent duct with a throat area of half the exit area is supplied with gas (k = 1.4) at stagnation pressure of 140 kPa. It discharges into an atmosphere of static pressure 100 kPa. Show that there is a normal shock in the duct. Find the stagnation pressure of the gas at exit from the nozzle, the Mach numbers immediately upstream and downstream of the shock, and the ratio of the area at the location of the shock to the throat area.

Solution

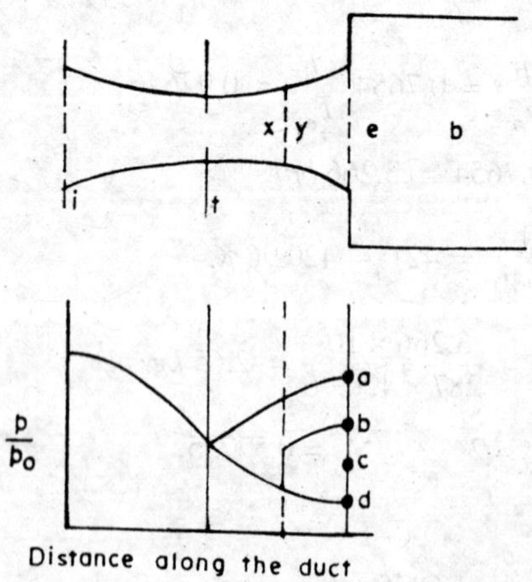

Fig. 6.20 Convergent divergent duct P.6.16

For isentropic choked flow.

$$\left(\frac{\dot{m}\ \sqrt{c_p T_o}}{A\,p_o}\right)_e = \left(\frac{\dot{m}\ \sqrt{c_p T_o}}{A\,p_o}\right)_t \times \frac{A_t}{A_e}$$

From isentropic flow table

$$\left(\frac{\dot{m}\ \sqrt{c_p T_o}}{A\,p_o}\right)_t = 1.281$$

$$\therefore \quad \left(\frac{\dot{m}\ \sqrt{c_p T_o}}{A\,p_o}\right)_e = 0.64 = \frac{p}{p_o}$$

For the subsonic exit (Point a)

$$\frac{p_e}{p_o} = 0.937,\ p_e = 0.937 \times 140 = 131\ kN/m^2 \text{ (too high)}$$

For supersonic exit (Point d)

$$\frac{p_e}{p_o} = 0.0938,\ \therefore\ p_e = 0.0938 \times 140 = 13.12\ kN/m^2$$

(too law)

Therefore there is a normal shock between the throat and the exit plane.

$$\frac{\dot{m}\ \sqrt{c_p T_o}}{A_e\,p_e} = \frac{\dot{m}\ \sqrt{c_p T_o}}{A_t\,p_o} \times \frac{A_t}{A_e} \times \frac{p_o}{p_e}$$

$$= 1.281 \times \frac{1}{2} \times 1.4 = 0.896$$

From the table $M_E = 0.4$ and $\dfrac{p_e}{p_{oe}} = 0.896$

[Note :— $\dfrac{\dot{m}\ \sqrt{c_p T_o}}{A\,p}$ is not tabulated but can be found by dividing $\dfrac{\dot{m}\ \sqrt{c_p T_o}}{A\,p_o}$ by p/p_o]

$$p_{oe} = \frac{100}{0.896} = 111.5\ kN/m^2$$

Total pressure ratio across the shock $= \dfrac{111.5}{140} = 0.796$

From the tables for normal shocks.

$$M_x = 1.84 \text{ and } M_y = 0.608$$

From the isentropic flow tables.

$$\frac{\dot{m}\sqrt{c_p T_o}}{A_x p_o} = 0.863$$

$$\frac{A_x}{A_t} = \left(\frac{\dot{m}\sqrt{c_p T_o}}{A p_o}\right)_t \left(\frac{\dot{m}\sqrt{c_p T_o}}{A p_o}\right)_x$$

$$= \frac{1.281}{0.863} = 1.49 \quad \text{Ans}$$

6.17 A convergent-divergent nozzle is designed for a pressure ratio of 0.12 $\left(\frac{p_e}{p_o} = 0.12\right)$ (a) Find the exhaust pressure which will locate the plane of the shock at the exit section. (b) Find the location of the shock when the pressure ratio is 0.6 (Assume that the nozzle has uniform divergence in its diverging section and that k = 1.4)

Solution

(a) From the isentropic tables, the ideal exit Mach number (Just ahead of the exit shock) is

At $\frac{p_x}{p_o} = 0.12, M_2 = M_x = 2.04$

From the shock tables.

At $M_x = 2.04, M_y = 0.57068$

$$\frac{p_y}{p_x} = 4.6886$$

Therefore, the exhaust pressure for a shock at the exit is

$$p_y = 4.6886\, p_x$$

$$= 4.6886 \times 0.12\, p_o = 0.5625\, p_o$$

(b) $$\frac{p_2}{p_{o1}} = \frac{p_2}{p_{o2}} \times \frac{p_{o2}}{p_{o1}} = \left(1 + \frac{k-1}{2} M_2^2\right)^{k/k-1} e^{-(\Delta s/R)}$$

$$\frac{A_2}{A_1^*} = \frac{1}{M_2}\left[\frac{1 + \frac{k-1}{2} M_2^2}{\frac{k+1}{2}}\right]^{\frac{k+1}{2(k-1)}} e^{\frac{\Delta S}{R}}$$

$$\therefore \quad \frac{p_2}{p_{o1}} \times \frac{A_2}{A_1^*} = \frac{1}{M_2}\left(\frac{2}{k+1}\right)^{\frac{k+1}{2(k-1)}}\left(1+\frac{k-1}{2}M_2^2\right)^{-\frac{1}{2}}$$

This equation is tabulated in the isentropic tables, but may be used here since the loss across the shock has been made to dropout.

From the isentropic tables.

$$\frac{A_2}{A_1^*} = 1.7452$$

Therefore, $\dfrac{p_2}{p_{o1}} \times \dfrac{A_2}{A_1^*} = (0.6)\,(1.7452) = 1.04712$

Since $\quad p_{o1} A_1^* = p_{o2} A_2^*$

$$\frac{p_2 A_2}{p_{o2} A_2^*} = 1.04712$$

From the isentropic tables

$M_2 = 0.537$

and $\quad \dfrac{p_2}{p_{o2}} = \dfrac{p_2}{p_{oy}} = 0.8218$

$$\frac{p_{oy}}{p_{ox}} = \frac{p_2/p_{ox}}{p_2/p_{oy}} = \frac{0.6}{0.8218} = 0.73$$

From the shock tabes, the Mach number ahead of the shock.

$M_x = 1.98$

$M_y = 0.5808$

The area ratio for isentropic flow from the throat to the shock location.

$$\frac{A_x}{A_1^*} = 1.6597 \qquad \text{Ans}$$

6.18 A nozzle is designed assuming reversible adiabatic flow with an exit Mach number of 2.6 while flowing air having a stagnation pressure and temperature of 2 MPa and 150°C, respectively. The mass rate of flow is 5 kg/s, and k may be assumed to 1.4 and constant.

(a) Determine the exit pressure, temperature and area and the throat area.

(b) Suppose that the back pressure at the nozzle exit is raised to 1.35

MPa, and that the flow remains isentropic except for a normal shock wave. Determine the exit Mach number and temperature and the mass flow through the nozzle.

Solution

(a) From isentropic tables

At $M_e = 2.6, \frac{p_e}{p_o} = 0.0501, \frac{A_e}{A^*} = 2.896$

Therefore, $p_e = 0.0501 \times 2 = 0.1002\ MPa$

From isentropic tables, at M = 1

$$\frac{T^*}{T_o} = 0.8333, T^* = 0.833 \times 423 = 352.486\ K$$

$$\frac{p^*}{T_o} = 0.5283,\ p^* = 2 \times 0.5283 = 1.0566\ MPa$$

$$a^* = \sqrt{k\,RT^*} = \sqrt{1.4 \times 287 \times 352.486} = 376.336\ m/s$$

$$V^* = M\,a^* = 1 \times a^* = 376.336$$

$$p^* = \rho^*\,RT^*$$

$$1.0566 \times 10^3 = \rho^* \times 0.287 \times 352.486$$

$$\rho^* = 10.44\ kg/m^3$$

$$\dot{m} = \rho^*\,A^*\,V^*$$

$$5 = 10.44 \times A^* \times 376.336$$

$$A^* = 1.272 \times 10^{-3}\ m^2 = 1272\ mm^2$$

From isentropic table at $M_e = 2.6$

$$\frac{A_e}{A^*} = 2.896,\ \frac{T_e}{T_o} = 0.425$$

Therefore,

$$A_e = 2.896 \times A^* = 2.896 \times 1.272 \times 10^{-3} = 3.68 \times 10^{-3}\ m^2$$

$$= 3684\ mm^2$$

$$\frac{T_e}{T_o} = 0.425$$

or, $T_e = 0.425 \times 423 = 179.775\ K$

(b) Assuming $M_x = 2$, $M_y = 0.57735$,

$$\frac{A_e}{A_x^*} = 2.896 \quad \frac{p_{oy}}{p_{ox}} = 0.721$$

At $M_x = 2$, $\dfrac{A_x}{A_x^*} = 1.6875$

and at $M_y = 0.57735$, $\dfrac{A_x}{A_y^*} = 1.2225$

Therefore, $\dfrac{A_e}{A_y^*} = \dfrac{A_e}{A_x^*} \times \dfrac{A_x^*}{A_y^*}$

$$= \frac{A_e}{A_x^*} \times \frac{A_x}{A_y^*} \times \frac{A_x^*}{A_x}$$

$$= 2.896 \times 1.2225 \times \frac{1}{1.6875} = 2.09779$$

Therefore, $M_e = 0.293$ *at* $A_e/A_y^* = 2.09799$

$$\frac{p_e}{p_{oy}} = 0.94171$$

$$\frac{p_e}{p_{ox}} = \frac{p_e}{p_{oy}} \times \frac{p_{oy}}{p_{ox}}$$

or, $\dfrac{p_e}{p_{ox}} = 0.94171 \times 0.721 = 0.67897$

Therefore, $p_e = 0.67897 \times p_{ox} = 0.67897 \times 20$

$= 13.58$ bar ≈ 1.35 MPa

Hence the assumed Mach number (Me = 0.293) is okay. Therefore,

$$\frac{T_e}{T_o} = 0.98298$$

$T_e = 0.98298 \times 423 = 415.8\ K$ **Ans**

$\dot{m} = 5\ kg/s$ **Ans**

6.19 A converging-diverging nozzle is designed to expand air isentropically to atmospheric pressure from a large tank where the temperature and absolute pressure are 150°C and 790 kPa. A normal shock stands in the diverging section at a location in the diverging section at a location where the pressure is 160 kPa and the cross sectional area is 600 mm^2. Determine the nozzle back pressure, exit area, and throat area.

Solution

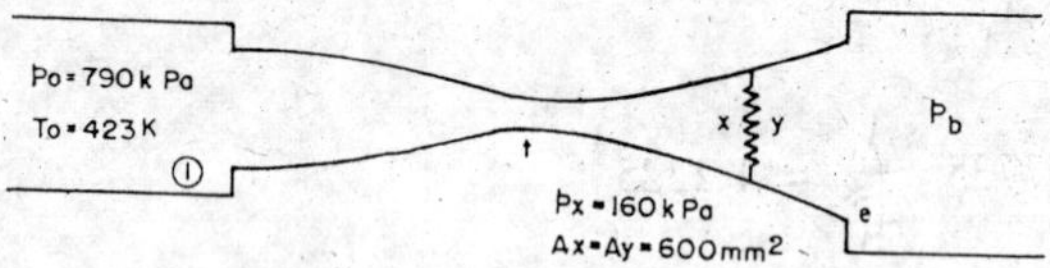

Fig. 6.21. converging-diverging nozzle P6.19

At design conditions

$$p_e/p_o = \frac{101}{790} = 0.1278$$

From isentropic tables

At $p_e/p_o = 0.1278;\ M_e = 2$ and $\dfrac{A_e}{A_t} = 1.687$

At section x, $\dfrac{p_x}{p_o} = \dfrac{160}{790} = 0.2025$

From isentropic tables

At $\dfrac{p_x}{p_o} = 0.2025,\ M_x = 1.7,\ \dfrac{A_x}{A_x^*} = 1.338$

Therefore, $A_t = A_x^* = \dfrac{A_x}{1.338} = \dfrac{600}{1.338} = 448.43\ mm^2$ Ans

Then $A_e = 1.687\,A_t = 1.687 \times 448.43 = 756.5\ mm^2$ Ans

From normal shock tables

At $M_x = 1.7,\ M_y = 0.641,$

$$\frac{p_{oy}}{p_{ox}} = 0.856,\ p_y/p_x = 3.205$$

Therefore, $p_{oy} = 0.856 \times 790 = 676.24\ kPa$

$$p_y = 3.205 \times 160 = 512.8\ kPa$$

From isentropic tables, at $M_y = 0.641$

$$\frac{A_y}{A_y^*} = 1.145$$

$$\text{Therefore, } A_y^* = \frac{A_y}{1.145} = \frac{600}{1.145} = 524\ mm^2$$

$$\text{Therefore, } \frac{A_e}{A_y^*} = \frac{756.5}{524} = 1.4437$$

At $\quad A_e/A_y^* = 1.4437,\ M_e \approx 0.45$

$$p_e/p_{oe} = 0.87$$

$$\text{Therefore, } p_b - p_e = 0.87 \times p_{oe} = 0.87 \times p_{oy}$$

$$= 0.87 \times 676.24 = 588.33 \text{ kPa}$$

Ans

6.20 Air moves through a converging diverging nozzle with a stagnation pressure of 800 kPa and a stagnation temperature of 375 K. The exit area ratio of the nozzle is 2. Assuming a normal shock at a section where Mach number is 1.8, determine the pressure, temperature, and Mach number downstream just after the shock and at the exit plane. Assume isentropic flow through the nozzle, except for irreversible compression in the normal shock.

Solution

Using isentropic tables.

At $\quad M_x = 1.8,\ \frac{p_x}{p_{ox}} = 0.174,\ \frac{T_x}{T_{ox}} = 0.607,\ \frac{A_2}{A_x^*} = 1.439$

$$\text{Therefore, } p_x = 800 \times 0.174 = 139.2\ kPa$$

$$T_x = 375 \times 0.607 = 227.625$$

Using shock tables

At $\quad M_x = 1.8,\ M_y = 0.6165$

$$\frac{p_y}{p_x} = 3.613,\ \frac{T_y}{T_x} = 1.532,\ p_o\frac{y}{p_{ox}} = 0.813$$

$$\text{Therefore, } p_y = 3.613 \times 139.2 = 502.9296\ kPa$$

$$T_y = 1.532 \times 227.625 = 348.7215\ K$$

$$p_{oy} = 0.813 \times 800 = 650\ kPa$$

It is to be noted that the stagnation temperature does not change across a normal shock

$$T_{ox} = T_{oy} = 375\ K$$

Using isentropic tables,

At $M_y = 0.6165$

$$\frac{p_y}{p_{oy}} = 0.774,\ \frac{T_y}{T_{oy}} = 0.929,\ \frac{A_y}{A_y^*} = 1.170$$

In the given problem $\frac{A_e}{A_x^*} = 2.$

Therefore, $p_y = 0.774 \times 650 = 503.1\ kPa$

$$T_y = 0.929 \times 375 = 348.375\ K$$

Since the flow from the point y to the exit plane e is isentropic, we may write

$$\frac{A_e}{A_e^*} = \frac{A_e}{A_y^*}$$

Therefore, $\frac{A_e}{A_y^*} = \frac{A_e}{A_x^*} \times \frac{A_x^*}{A_x} \times \frac{A_x}{A_y} \times \frac{A_y}{A_y^*}$

$$= 2 \times \frac{1}{1.439} \times 1 \times 1.170 = 1.626$$

From the isentropic tables,

At $\frac{A_e}{A_e^*} = \frac{A_e}{A_y^*} = 1.626$ and $M < 1$

$$M_e \approx 0.39,\ \frac{p_e}{p_{oy}} = 0.9,\ \frac{T_e}{T_{oy}} = 0.97$$

Therefore, $p_e = 0.9 \times 650 = 585\ kPa$ Ans

$T_e = 0.97 \times 375 = 363.75\ K$ Ans

6.21 The total temperature and total pressure of air in a reservoir supplying a converging-diverging nozzle are 450 K and 4 bar respectively. The nozzle throat is 6 cm^2 and the nozzle exit area is 18 cm^2. A shock is noted at a position in the diverging portion where the area is 12 cm^2.

(a) What are the exit pressure, temperature and velocity?

(b) What value of the back pressure would just cause the flow through the nozzle to be completely supersonic ?

(c) What value of back pressure would result in completely isentropic flow interior and enterior to the nozzle ?

Solution

$$\frac{A}{A^*} = \frac{12}{6} = 2$$

From isentropic tables

At $\quad A/A^* = 2, M_x = 2.2, \dfrac{p_x}{p_{ox}} = 0.09352, \dfrac{T_x}{T_{ox}} = 0.50813$

$$\frac{\rho_x}{\rho_{ox}} = 0.184$$

Therefore, $T_x = 0.50813 \times 450 = 228.65\ K$

$$p_x = 0.09352 \times 4 = 0.374\ bar$$

From shock tables

At $\quad M_x = 2.2, M_y = 0.547, \dfrac{p_y}{p_x} = 5.48,$

$$\frac{T_y}{T_x} = 0.62812, \frac{p_{oy}}{p_{ox}} = 0.62812$$

Therefore, $T_y = 0.62812 \times 228.65 = 143.62\ K$

$$p_y = 5.48 \times 0.374 = 2.05\ bar$$

$$p_{oy} = 0.62812 \times 4 = 2.512\ bar$$

Area ratio corresponding to shock

$$\frac{A_e}{A^*} = \frac{A_e}{A_x^*} \times \frac{A_x^*}{A_y^*} = \frac{18}{6} \times 0.62812 = 1.884$$

From isentropic tables

At $\quad \dfrac{A_e}{A^*} = 1.884, M_e = 0.33, \dfrac{p_e}{p_o} = 0.92736,$ Ans

$$\frac{T_e}{T_o} = 0.97868$$

Therefore, $p_e = 0.92736 \times 2.512 = 2.33\ bar$ Ans

$$T_e = 0.97868 \times 450 = 440.4\ K$$ Ans

$$a_e = \sqrt{k\,RT_e} = \sqrt{1.4 \times 287 \times 440.4} = 420.657$$

$$V_e = M_e\,a_e$$

$$= 0.33 \times 420.657 = 138.82$$ Ans

(b) When the flow is just supersonic then

$$p_b = p_y$$

From isentropic tables,

At $\dfrac{A}{A^*} = 3,\ M_x = 2.64$ and $\dfrac{p_x}{p_{ox}} = 0.04711$

From shock table

At $M_x = 2.64,\ M_y = 0.5$ and $\dfrac{p_y}{p_x} = 7.9645$

Therefore, $p_y = \dfrac{p_y}{p_x} \times \dfrac{p_x}{p_{ox}} \times p_{ox}$

$$= 7.964 \times 0.04711 \times 4 = 1.5 \text{ bar}$$ Ans

(c) When the complete flow is isentropic then $p_b = p_x$

At $\dfrac{A}{A^*} = 3,\ M_x = 2.64$ and $\dfrac{p_x}{p_{ox}} = 0.04711$

Therefore, $p_x = 0.04711 \times 4 = 0.18844\ bar = p_b$

6.22 Derive the relation for normal shock

$$M_y^2 = \frac{M_x^2 + 2/(k-1)}{\dfrac{2k}{k-1} M_x^2 - 1}$$

<u>Solution</u> We know the Rayleigh line equation in terms of p and M

$$\frac{p_y}{p_x} = \frac{1 + k\,M_x^2}{1 + k\,M_y^2} \tag{1}$$

We also know the Fanno line equation in terms of p and M

$$\frac{p_y}{p_x} = \frac{M_x \sqrt{1 + \dfrac{k-1}{2} M_x^2}}{M_y \sqrt{1 + \dfrac{k-1}{2} M_y^2}} \tag{2}$$

Equating equations (a) and (2)

$$\frac{M_x \sqrt{1+\frac{k-1}{2} M_x^2}}{1+k M_x^2} = \frac{M_y \sqrt{1+\frac{k-1}{2} M_y^2}}{1+k M_y^2}$$

Squaring on both sides and by cross multipication.

$$\left(M_x^2 + \frac{k-1}{2} M_x^4\right)\left(1+2k M_y^2 - k^2 M_y^4\right)$$

$$-\left(M_y^2 + \frac{k-1}{2} M_y^4\right)\left(1+2k M_x^2 + k^2 M_x^4\right) = 0$$

Opening the brackets and rearranging

$$M_x^2 - M_y^2 + k^2 M_x^2 M_y^2 \left(M_y^2 - M_x^2\right) + \frac{k-1}{2}\left(M_x^4 - M_y^4\right)$$

$$+ k(k-1) M_x^2 M_y^2 \left(M_x^2 - M_y^2\right) = 0$$

Eliminating common terms $\left(M_x^2 - M_y^2\right)$

$$1 - k M_x^2 M_y^2 + \frac{k-1}{2}\left(M_x^2 - M_y^2\right) + k(k-1) M_x^2 M_y^2 = 0$$

or, $2 - 2k^2 M_x^2 M_y^2 - (k+1) M_x^2 + (k-1) M_y^2$

$$+ 2k(k-1) M_x^2 M_y^2 = 0$$

or, $2 + (k-1) M_x^2 + (k-1) M_y^2 - 2k M_x^2 M_y^2 = 0$

Dividing by (k – 1)

$$M_x^2 + \frac{2}{k-1} = -M_y^2 + \left(\frac{2k}{k-1}\right) M_x^2 M_y^2$$

or, $M_y^2\left(\frac{2k}{k-1} M_x^2 - 1\right) = M_x^2 + \frac{2}{k-1}$

or, $M_y^2 = \dfrac{M_x^2 + 2/k-1}{\left(\frac{2k}{k-1}\right) M_x^2 - 1}$ Proved

6.23 Show that the stagnation pressure ratio across the normal shock is given by

$$\frac{p_{oy}}{p_{ox}} = \left[\frac{\frac{k+1}{2} M_x^2}{1+\frac{k-1}{2} M_x^2}\right]^{k/k-1} \left[\frac{2k M_x^2}{k+1} - \frac{k-1}{k+1}\right]^{1/k-1}$$

Solution

$$\frac{p_{oy}}{p_{ox}} = \frac{p_{oy}}{p_y} \times \frac{p_y}{p_x} \times \frac{p_x}{p_{ox}} \qquad (1)$$

We have

$$\frac{p_{ox}}{p_x} = \left(\frac{T_{ox}}{T_x}\right)^{k/k-1} = \left[1 + \frac{k-1}{2} M_x^2\right]^{k/k-1}$$

$$\frac{p_{oy}}{p_y} = \left(\frac{T_{oy}}{T_y}\right)^{k/k-1} = \left[1 + \frac{k-1}{2} M_y^2\right]^{k/k-1}$$

and $$\frac{p_y}{p_x} = \left(\frac{2k}{k+1}\right) M_x^2 - \frac{k-1}{k+1}$$

Substituting in equation (1)

$$\frac{p_{oy}}{p_{ox}} = \left(\frac{1 + \frac{k-1}{2} M_y^2}{1 + \frac{k-1}{2} M_x^2}\right)^{k/k-1} \left(\frac{2k}{k+1} M_x^2 - \frac{k-1}{k+1}\right) \qquad (2)$$

We know that

$$M_y^2 = \frac{M_x^2 + \frac{2}{k-1}}{\frac{2k}{k-1} M_x^2 - 1}$$

Substituting in equation (2)

$$\frac{p_{oy}}{p_{ox}} = \left[1 + \frac{\left(\frac{k-1}{2}\right)\left(M_x^2 + \frac{2}{k-1}\right)}{\frac{2k}{k-1} M_x^2 - 1}\right]^{k/k-1} \frac{\left(\frac{2k}{k+1} M_x^2 - \frac{k-1}{k+1}\right)}{\left(1 + \frac{k-1}{2} M_x^2\right)^{k/k-1}}$$

Simplifying the term under square brackets

$$\left[\frac{\frac{2k}{k-1} M_x^2 - 1 + \frac{k-1}{2} M_x^2 + 1}{\frac{2k}{k-1} M_x^2 - 1}\right]^{k/k-1}$$

$$= \left[\frac{\frac{(k+1)^2}{2(k-1)} M_x^2}{\frac{2k}{k-1} M_x^2 - 1}\right]^{k/k-1}$$

$$= \left[\frac{\left(\frac{k+1}{2} M_x^2\right)\left(\frac{k+1}{k-1}\right)}{\left(\frac{2k}{k-1} M_x^2\right)\left(\frac{k+1}{k-1}\right) - 1} \right]^{k/k-1}$$

$$= \left[\frac{\left(\frac{k+1}{2} M_x^2\right)\left(\frac{k+1}{k-1}\right)}{\left(\frac{2k}{k-1} M_x^2\right)\left(\frac{k+1}{k-1}\right) - 1} \right]^{k/k-1}$$

$$= \left[\frac{\left[\frac{k+1}{2} M_x^2\right]\left[\frac{k+1}{k-1}\right]}{\left[\frac{2k}{k-1} M_x^2 - \frac{k-1}{k+1}\right]^{k/k-1} \left(\frac{k+1}{k-1}\right)} \right]^{k/k-1}$$

$$= \left[\frac{\left(\frac{k+1}{2} M_x^2\right)}{\left(\frac{2k}{k-1} M_x^2 - \frac{k-1}{k+1}\right)} \right]^{k/k-1}$$

$$\frac{p_{oy}}{p_{ox}} = \left[\frac{\left(\frac{k+1}{2} M_x^2\right)}{\left(\frac{2k}{k-1} M_x^2 - \frac{k-1}{k+1}\right)} \right]^{k/k-1} \frac{\left(\frac{2k}{k+1} M_x^2 - \frac{k-1}{k+1}\right)}{\left(1 + \frac{k-1}{2} M_x^2\right)^{k/k-1}}$$

$$= \left[\frac{\frac{k+1}{2} M_x^2}{1 + \frac{k-1}{2} M_x^2} \right]^{k/k-1} \left(\frac{2k\,M_x^2}{k+1} - \frac{k-1}{k+1}\right)^{1/k-1}$$

Proved

6.24 Show that for air (k = 1.4) the increase in density due to a normal shock cannot be greater six times the initial density.

Also prove that the corresponding minimum value of the Mach number after normal shock is $\sqrt{(k-1)/2k}$

Solution

We know that for normal shock

$$\frac{\rho_y}{\rho_x} = \frac{(k+1)\,M_x^2}{2 + (k-1)\,M_x^2}$$

Differentiating with respect to M_x.

$$\frac{d(\rho_y/\rho_x)}{dM_x}$$

$$= \frac{[2+(k-1)M_x^2][2M_x(k+1)]-(k+1)M_x^2-2M_x(k-1)}{[2+(k-1)M_x^2]^2}$$

For maximum value of ρ_y/ρ_x first derivative should be zero

Therefore, $d(\rho_y/\rho_x)/dM_x = 0$

or, $$\frac{2M_x(k+1)(2+kM_x^2-M_x^2-kM_x^2+M_x^2)}{[2+(k-1)M_x^2]^2} = 0$$

or, $$\frac{4M_x(k+1)}{[2+(k-1)M_x^2]^2} = 0$$

This gives the value $M_x = \infty$

Therefore the maximum value of ρ_y/ρ_x is

$$\left(\frac{\rho_y}{\rho_x}\right)_{max} = \frac{(k+1)M_x^2}{2+(k-1)M_x^2}$$

Dividig by M_x

Therefore,

$$\left(\rho_y/\rho_x\right)_{Max} = \frac{(k+1)}{\frac{2}{M_x^2}+(k-1)}$$

Putting the value of $M_x = \infty$ and k = 1.4

$$\left(\rho_y/\rho_x\right)_{Max} = \frac{1.4+1}{0+(1.4-1)} = \frac{2.4}{0.4} = 6$$

Minimum value of the Mach number after the shock is

$$M_y = \sqrt{\left[\frac{2/k-1+M_x^2}{[2k/(k-1)]\,M_x^2-1}\right]}$$

Dividing by M_x

$$\text{Therefore, } M_y = \sqrt{\left[\frac{\frac{2}{(k-1)M_x^2}+1}{\frac{2k}{k-1}-\frac{1}{M_x^2}}\right]}$$

Putting $M_x = \infty$

Therefore,

$$M_{y_{min}} = \sqrt{\left[\frac{0+1}{\frac{2k}{k-1}}\right]}$$

$$= \sqrt{\frac{k-1}{2k}}$$ Proved

6.25 (a) Define the following terms

Weak shock wave

Strength of shock wave

Shock of vanishing strength

Irreversibility of weak shock

(b) A supersonic diffuser for air has an area ratio of 0.416 with an inlet Mach number of 2.4 (design value). Determine the exit Mach number and the design value of the pressure ratio across the diffuser for isentropic flow. At an off design value of the inlet Mach number of 2.7 a normal shock occurs inside the diffuser. Determine the upstream Mach number and area ratio at the section where the shock occurs, diffuser efficiency and pressure ratio across the diffuser.

Solution

(a) See text

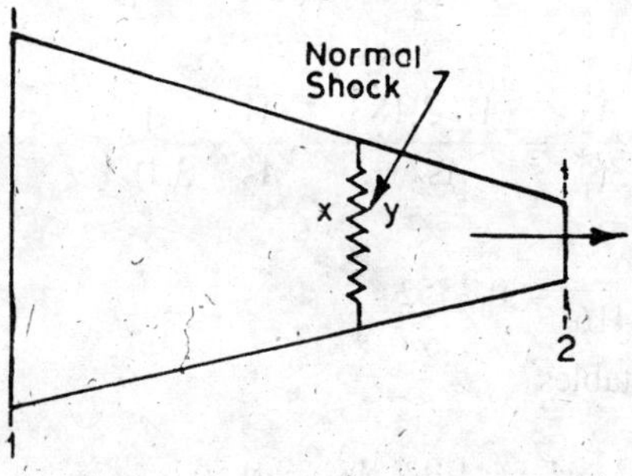

Fig. 6.22 Supersonic diffuser with normal shock P6.25

(b)

$$\frac{A_2}{A_1} = \frac{A_2}{A_2^*} \times \frac{A_1^*}{A_x} = 0.416$$

From isentropic tables

At $M_1 = 2.4, \frac{A_1}{A_1^*} = 2.403, \frac{p_1}{p_o} = 0.0684$

Therefore, $\frac{A_2}{A_2^*} = 0.416 \times 2.403 \approx 1$

From isentropic tables

At $\frac{A_2}{A_2^*} = 1, M_2 = 1, \frac{p_2}{p_o} = 0.528$ Ans

Therefore,

$$\frac{p_2}{p_1} = \frac{p_2}{p_o} \times \frac{p_o}{p_1} = \frac{0.528}{0.0684} = 7.7193$$ Ans

Off design values

From isentropic tables

At $M_1 = 2.7, \frac{p_1}{p_{o1}} = 0.043, \frac{T_1}{T_{o1}} = 0.407$

$$\frac{A_1}{A_1^*} = \frac{A_1}{A_x^*} = 3.183$$

and at $M_2 = 1, \frac{p_2}{p_{o2}} = 0.528, \frac{A_2}{A_2^*} = \frac{A_2}{A_2^*} = 1$

$$\text{Now, } \frac{p_{oy}}{p_{ox}} = \frac{p_{o2}}{p_{o1}} = \frac{A_2^*}{A_1^*} = \frac{A_1/3.183}{A_2/1} = \frac{A_1}{A_2} \times \frac{1}{3.183}$$

$$= \frac{1}{3.183 \times 0.416} = 0.7552$$

From normal shock tables

At $\frac{p_{oy}}{p_{ox}} = 0.7552, M_x = 1.92, M_y = 0.592$

$A_x/A_x^* = 1.58$

$M_x = 1.92$ Ans

$$\frac{A_x}{A_1} = \frac{A_x}{A_x^*} \times \frac{A_1^*}{A_1} = \frac{1.58}{3.183} = 0.49638$$ Ans

The diffuser efficiency at off design with shock is given by

$$\eta_D = \frac{\dfrac{T_{o1}}{T_1}\left(\dfrac{p_{oy}}{p_{ox}}\right)^{k-1/k} - 1}{\dfrac{k-1}{2} M_1^2} \times 100\ \%$$

$$= \frac{\dfrac{1}{0.407}(0.7552)^{0.286} - 1}{0.2\,(2.7)^2} \times 100\ \%$$

$$= \frac{2.2674 - 1}{1.458} \times 100\ \%$$

$$= \frac{1.2674}{1.458} \times 100\ \%$$

$$= 0.86928 \times 100\ \% = 86.928\ \%$$ Ans

6.26 Consider the diffuser of a supersonic aircraft flying at M = 1.4 at such an altitude that the temperature is – 20°C, and the atmospheric pressure is 50 kPa. Consider two possible ways in which this diffuser might operate, and for each case calculate the throat area required for a flow of 50 kg/s.

(a) The diffuser operates as a reversible adiabatic diffuser with subsonic exit velocity.

(b) A normal shock stands at the entrance to the diffuser. Except for the normal shock the flow is reversible and adiabatic, and the exit velocity is subsonic. This is shown in Fig 6.23. Assume a convergent divergent diffuser with M = 1 at the throat.

Solution

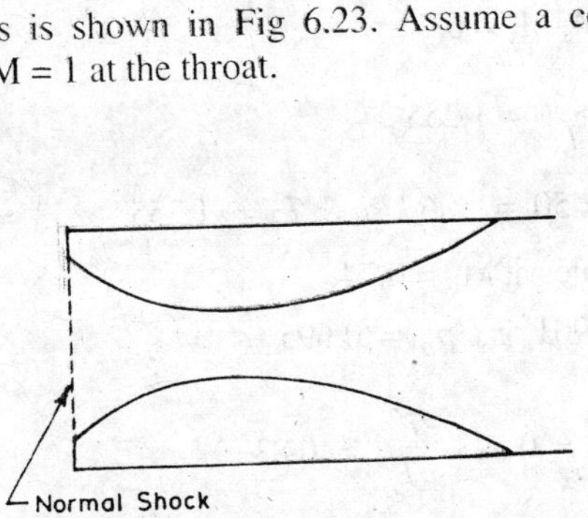

Fig. 6.23 Diffuser with normal shock P6.26

From isentropic tables.

At $\quad M = 1.4, \quad \dfrac{p_1}{p_{o1}} = 0.314, \quad \dfrac{T_1}{T_{o1}} = 0.718$

and at $M = 1$, $\dfrac{p^*}{p_{o1}} = 0.5283$, $\dfrac{T^*}{T_{o1}} = 0.8333$

Therefore, $p_{o1} = \dfrac{p_1}{0.314} = \dfrac{50}{0.314} = 159.2356$

$$p^* = 0.5283 \times 159.2356 = 84.12\ kPa$$

$$T_{o1} = \frac{T_1}{0.718} = \frac{253}{0.718} = 352.367$$

$$T^* = 0.8333 \times 352.367 = 293.63\ K$$

$$a^* = \sqrt{kRT^*} = \sqrt{1.4 \times 287 \times 293.63} = 343.48\ m/s = V^*$$

$$p^* = \rho^* RT^*$$

Therefore, $\rho^* = \dfrac{p^*}{RT^*} = \dfrac{84.12}{0.287 \times 293.63} = 0.998\ kg/m^3$

$$\dot{m} = \rho^* A^* V^*$$

$$50 = 0.998 \times A^* \times 343.48$$

$$A^* = 0.1458\ m^2 = 1458\ mm^2$$ Ans

(b) At $M_x = 1.4$, $M_y = 0.74$

$$\frac{p_y}{p_x} = 2.12, \quad \frac{T_y}{T_x} = 1.255$$

$$p_y = 2.12 \times 50 = 106\ kPa\ \&\ T_y = 1.255 \times 253 = 317.5\ K$$

From isentropic tables at $M_y = 0.74$

$$T_y/T_o = 0.901, \quad p_y/p_o = 0.695$$

and at $M = 1$, $\dfrac{p^*}{p_o} = 0.528$, $\dfrac{T^*}{T_o} = 0.833$

Therefore, $T_o = \dfrac{T_y}{0.901} = \dfrac{317.5}{0.901} = 352.4$

and $T^* = 0.833 \times 352.4 = 293.65\ K$

$$p_o = \frac{p_y}{0.695} = \frac{106}{0.695} = 152.5\ kPa$$

$$p^* = 0.528 \times p_o = 0.528 \times 152.5 = 80.53\ kPa$$

$$a^* = \sqrt{k\,RT^*} = \sqrt{1.4 \times 287 \times 293.65} = 343.5\ m/s = V^*$$

$$p^* = \rho^*\,RT^*$$

$$\rho^* = \frac{p^*}{RT^*} = \frac{80.53}{0.287 \times 293.65} = 0.955$$

Therefore $\dot{m} = \rho^*\,A^*\,V^*$

$$50 = 0.955 \times A^* \times 343.5$$

$$A^* = 0.1523\ m^2$$ Ans

6.27 (a) Derive the Rayleigh supersonic pitot tube formula

$$\frac{p_{oy}}{p_{ox}} = \left(\frac{k+1}{2}M_x^2\right)^{k/k-1}\left(\frac{2k}{k+1}M_x^2 - \frac{k-1}{k+1}\right)^{1/k-1}$$

and plot the curve between M_x and p_{oy}/p_x.

(b) The ratio of the entry to exit area in a subsonic diffuser is 0.25. The Mach number of a jet of air approaching the diffuser at standard condition (p_o = 101.3 kPa, T_o = 290 K is 2.2. There is a standing normal shock wave just outside the diffuser is isentropic. Determine at the exit of the diffuser Mach number, temperature and pressure. What is the stagnation pressure loss between the initial and final states of the flow ?

Solution

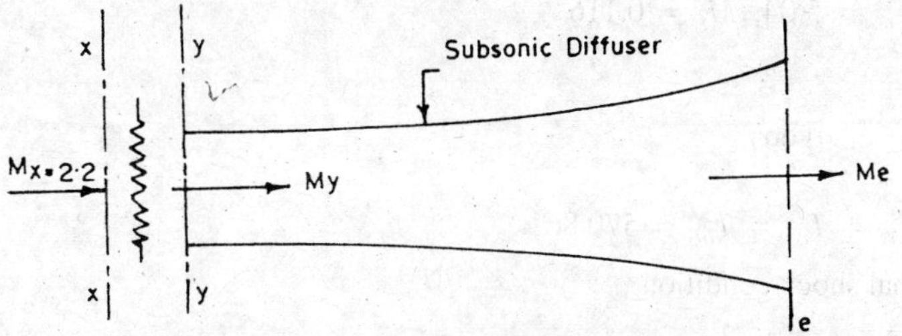

Fig. 6.24. Normal Shock at diffuser entry P 6.27

(a) See text

(b) From isentropic tables

At $M_x = 2.2, \dfrac{p_x}{p_{ox}} = 0.0935, \dfrac{T_x}{T_{ox}} = 0.508$

Therefore, $p_x = 0.0935 \times 101.3 = 9.47\ kPa$

$$T_{ox} = \frac{290}{0.508} = 570.86\ K$$

From normal shock tables

At $M_x = 2.2, M_y = 0.547, \dfrac{p_y}{p_x} = 5.48,$

$$T_y/T_x = 1.857, \frac{p_{oy}}{p_{ox}} = 0.628$$

Therefore, $p_y = 5.48 \times 9.47 = 51.9\ kPa$

$$T_y = 1.857 \times 290 = 538.53\ kPa$$

From isentropic tables

At $M_y = 0.547, \dfrac{A_y}{A_y^*} = 1.26 = \dfrac{A_y}{A_e^*}$

Therefore, $\dfrac{A_y}{A_y^*} = \dfrac{A_y}{A_e^*}$

$$\frac{A_e}{A_e^*} = \frac{A_y}{A_e^*} \times \frac{A_e}{A_y} = 1.26 \times 4 = 5.04$$

From isentropic tables

At $\dfrac{A_e}{A_e^*} = 5.04, M_e = 0.116$ Ans

$$\frac{T_e}{T_{oe}} = 0.997$$

and $T_{ox} = T_{oy} = T_{oe} = 570.86$

For normal shock condition

Therefore,

$$T_e = 0.997 \times 570.86 = 569.147\ K$$ Ans

From isentropic tables

At $M_y = 0.547, \dfrac{p_y}{p_{oy}} = 0.816$

$$\text{Therefore, } p_{oy} = \frac{p_y}{0.816} = \frac{51.9}{0.816} = 63.6\ kPa = p_{oe}$$

From isentropic tables

At $\quad M_e = 0.116$

$$\frac{p_e}{p_{oe}} = 0.99$$

Therefore, $p_e = 1.099 \times 63.6 = 62.96\ kPa$ Ans

The stagnation pressure loss

$$p_{ox} - p_{oy} = 101.3 - 63.6 = 37.7\ kPa$$ Ans

6.28 A supersonic wind tunnel has a working section 2.5 × 2.5 m. If the Mach number in the working section is to be 2.2, find the area of the throat of the nozzle. Neglecting all losses except due to essential normal shocks, estimate the size of throat of diffuser section to start the wind tunnel. Assume that stagnation pressure and temperature at the inlet are 100 kPa and 290 K and that geometry of wind tunnel is not varied while the tunnel is running. Calculate the pressure drop across the tunnel.

Solution

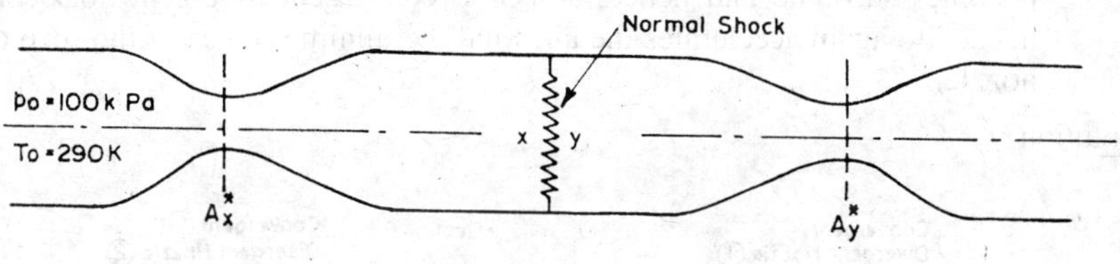

Fig. 6.25 Supersonic wind tunnel P6.28

$$A_x = 2.5 \times 2.5 = 6.25\ m^2$$

$$M_x = 2.2$$

From isentropic tables

At $\quad M_x = 2.2, \quad \frac{A_x}{A_x^*} = 2.005,$

$$\text{Therefore, } A_x^* = \frac{6.25}{2.0005} = 3.117\ m^2$$

From normal shock tables

At $M_x = 2.2, \dfrac{p_{oy}}{p_{ox}} = 0.628$

Therefore, $p_{oy} = 0.628 \times 100 = 62.8\ kPa$

Also, $\dfrac{A_x}{A_y^*} = \dfrac{A_x}{A_x^*} \times \dfrac{p_{oy}}{p_{ox}}$

$= 2.005 \times 0.628 = 1.259$

$A_y^* = \dfrac{A_x}{1.259} = \dfrac{6.25}{1.259} = 4.96\ m^2$

Diffuser section throat area $A_y^* = 4.96\ m^2$ Ans

Pressure drop across the wind tunnel

$p_{oy} - p_{ox}$

$= 100 - 62.8 = 37.2\ kPa$ Ans

6.29 A convergent-divergent nozzle having a minimum area of 0.05 m^2, accelerates air to a supersonic velocity and supplies it to a duct to cross sectional area 0.1 m^2. However, due to a shock in this section the velocity becomes subsonic and hence, a second convergent divergent nozzle is needed to again accelerates the air. Find the minimum area of the second nozzle.

Solution

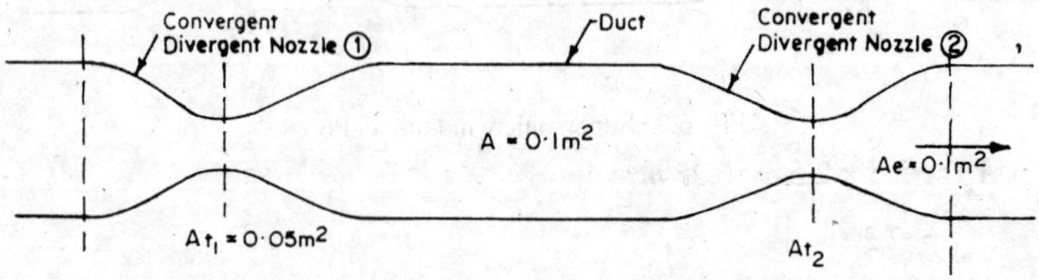

Fig. 6.26 Convergent divergnent nozzles and duct P6.29

At exit of first nozzle $\dfrac{A_1}{A^*} = \dfrac{0.1}{0.05} = 2$

Using the isentropic tables, the Mach number lies between 2.19 and 2.2. Using interpolation the Mach number comes out to be 2.197.

The value of M_2 can be calculated with respect to $M_1 = 2.197$. Using normal shock tables and doing interpolation, $M_2 = 0.5475$.

Using isentropic tables, by interpolation, at

$$M_2 = 0.5475. \frac{A_{t_2}}{A^*} = 1.258$$

$$\frac{A_e}{A^*} = 1.258$$

or, $$A^* = \frac{A_e}{1.258} = \frac{0.1}{1.258} = 0.0795\ m^2$$ Ans

6.30 (a) Derive the Rankine-Hugoniot relationship for a normal shock and compare them with isentropic relations.

(b) A normal shock occurs in the diverging section of a convergent divergent air nozzle. The exit area is 3 times the throat area and the stagnation pressure at the entry is 2.5 times the static pressure at exit. The flow is throughout isentropic except through the shock. Determine

(a) the Mach number M_x and M_y

(b) the static pressure and

(c) the area of cross section of the nozzle at the section of the nozzle where the normal shock occurs.

Solution

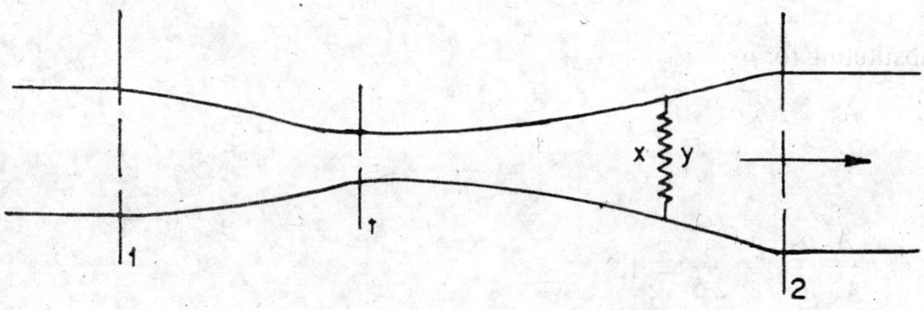

Fig. 6.27. Convergent-divergent nozzle P 6.30

(a) See text

(b)

We know that

$$\frac{\dot{m}\ \sqrt{T_o}}{A^*\ p_o} = Constant$$

For a normal shock $T_{ox} = T_{oy} = T_o$

Therefore,

$$A^* p_o = Constant$$

$$A_1^* p_{o1} = A_2^* p_{o2} \quad or, \quad \frac{A_1^*}{A_2^*} = \frac{p_{o2}}{p_{o1}}$$

Hence, $p_{o1} = p_{ot} = p_{ox}$

$$p_{o2} = p_{oy} \text{ and } A_1^* = A_t = A_t^* = A_2^*$$

$$A_y^* = A_2^*$$

Therefore, $\dfrac{A_1}{A_2^*} = \dfrac{A_1}{A_1^*} \times \dfrac{A_1^*}{A_2^*} = \dfrac{A_1}{A_1^*}\dfrac{p_{o2}}{p_{o1}}$

Similarly $\dfrac{A_t}{A_2^*} = \dfrac{A_t}{A_t^*} \times \dfrac{p_{o2}}{p_{o1}} = \dfrac{p_{o2}}{p_{o1}}$

Now $\dfrac{A_2}{A_t} = \dfrac{A_2}{A_2^*} \times \dfrac{A_2^*}{A_t} = \dfrac{A_2}{A_2^*} \times \dfrac{p_{o1}}{p_{o2}} = 3$

Substituting for p_{o1}

$$\frac{A_2}{A_2^*} \times \frac{2.5\, p_2}{p_{o2}} = 3$$

$$\frac{A_2 p_2}{A_2^* p_{o2}} = \frac{3}{2.5} = 1.2$$

From isentropic tables

At $\dfrac{A}{A^*}\left(\dfrac{p}{p_o}\right) = 1.2, \; M = 0.472 = M_2$

and $p/p_o = \dfrac{p_2}{p_{o2}} = 0.859$

$$\frac{p_2}{p_{o2}} = 0.859$$

$$p_{o2} = \frac{p_2}{0.859} = \frac{p_{o1}}{2.5 \times 0.859}$$

$$= \frac{p_{o1}}{2.1475} = 0.4656\, p_{o1}$$

Therefore, $\frac{p_{o2}}{p_{o1}} = 0.4656 = \frac{p_{oy}}{p_{ox}}$

(a) From normal shock tables

At $\frac{p_{oy}}{p_{ox}} = 0.4656,\ M_x = 2.58,\ M_y = 0.506$ Ans

$\frac{p_{oy}}{p_x} = 9.115$

(b) $p_x = \frac{p_{oy}}{9.115} = \frac{0.4656}{9.115} p_{ox} = 0.051\, p_{ox}$ Ans

Alternatively, From isentropic tables at

$M_x = 2.58,\ p/p_o = 0.051$ Ans

(c) From isentropic tables

At $M_x = 2.58$

$A/A^* = 2.842$

Therefore, $\frac{A_x}{A_x^*} = \frac{A_x}{A_t} = \frac{A}{A^*} = 2.842$

Therefore shock occurs in the diverging portion where $A_x = 2.842\ A_t$ Ans

6.31 Fuel is to be burned in a stream of air moving with a Mach number of 2, temperature 167 K and a pressure of 68 kPa in a duct of constant area. Find the maximum possible rise in stagnation temperature and corresponding values of temperature, pressure in stagnation pressure.

If the supersonic stream given by the conditions above were to pass through a normal shock before fuel was burned; find the maximum possible rise in temperature and corresponding values of temperature, pressure and stagnation pressure.

Solution For the maximum heat transfer, $M_2 = 1$.

From isentropic tables

At $M_1 = 2,\ \frac{p_1}{p_{o1}} = 0.128,\ \frac{T_1}{T_{o1}} = 0.555$

Therefore, $p_{o1} = \dfrac{618}{0.128} = 531.25\ kPa$

and $T_{o1} = \dfrac{167}{0.555} = 301\ K$

Also from Rayleigh tables

At $M_1 = 2, \dfrac{T_{o1}}{T_o^*} = 0.793, \dfrac{T_1}{T^*} = 0.529,$

$$p_1/p^* = 0.364 \text{ and } \frac{p_{o1}}{p_o^*} = 1.503$$

Therefore, $T_o^* = \dfrac{301}{0.793} = 379.57\ K$

$$T^* = \frac{167}{0.529} = 315.689\ K \qquad \text{Ans}$$

$$p_o^* = \frac{531.25}{1.5} = 354.157\ kPa \qquad \text{Ans}$$

$$p^* = \frac{68}{0.364} = 186.8\ kPa \qquad \text{Ans}$$

Maximum possible rise in stagnation temperature

$$T_{o2} - T_{o1} = T_o^* - T_{o1} = 379.57 - 301 = 78.57\ K \qquad \text{Ans}$$

(b) There is normal shock before fuel is burned and the flow is isentropic upto it.

From normal shock tables

At $M_x = 2,\ M_y = 0.5774, \dfrac{p_y}{p_x} = 4.5,$

$$\frac{T_y}{T_x} = 1.688, \frac{p_{oy}}{p_{ox}} = 0.721$$

Therefore,

$$p_y = 4.5 \times 68 = 306\ kPa$$

$$p_{oy} = 0.721 \times 531.25 = 383.03\ kPa$$

$$T_y = 1.688 \times 167 = 281.896\ K$$

$$T_{ox} = T_{oy} = 301\ K$$

Again for maximum heat transfer, $M_2 = 1$

At $M_y = 0.577,\ p_y/p^* = 1.632,\ \frac{T_{oy}}{T_{ox}} = 0.808$

$T_y/T^* = 0.889,\ \frac{p_{oy}}{p_{ox}} = 1.083$

Therefore, $p^* = \frac{306}{1.632} = 187.5\ kPa$

$T^* = \frac{281.896}{0.899} = 317.09\ K$

$T_o^* = 301/0.808 = 372.52\ K$

$p_o^* = \frac{383.03}{1.083} = 353.67\ kPa$

The results obtained for parts (a) and (b) are approximately same. (except for small variations due to calculations and three digit tables used) From the Rayleigh live we can see that the both the situations are the same.

6.32 An isentropic nozzle having an area ratio of 2 discharges air into an insulated pipe of length L and diameter D. The nozzle is supplied at 660 kPa and 325 K and the duct discharges into a space where the pressure is 272 kPa. Calculate λ_{av} L/D of the pipe and mass flow per unit area in the pipe when shock stands (a) at exit section of the nozzle (b) at the exit section of the duct.

Solution

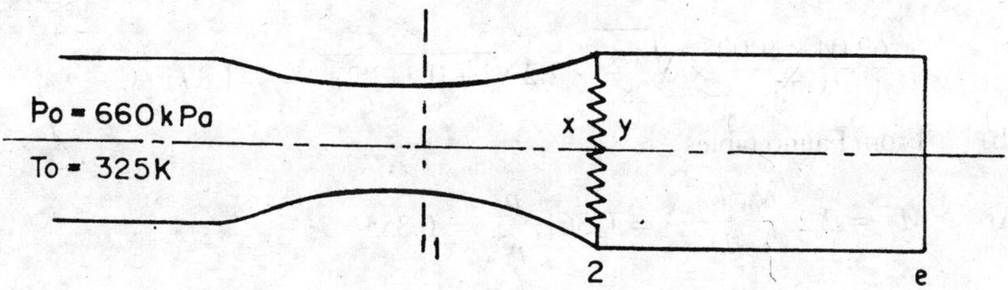

Fig. 6.28 Isentropic nozzle with insulated pipe P 6.32.

From isentropic tables

At $\frac{A}{A^*} \doteq 2,\ M = 2.2,\ \frac{p}{p_o} = 0.094$

Therefore, p = 0.094 × 660 = 62.04 kPa

As shock stands at exit section of the nozzle, from normal shock tables

At $\quad M_x = 2.2, M_y = 0.547, \dfrac{p_y}{p_x} = 5.48$

From Fanno tables

$$M_y = 0.547, \frac{\lambda_{av} L_{max}}{D} = 0.747, \; p_y/p^* = 1.94$$

Also, $\dfrac{p_e}{p^*} = \dfrac{p_y}{p^*} \times \dfrac{p_x}{p_y} \times \dfrac{p_o}{p_x} \times \dfrac{p_e}{p_o}$

Therefore, $\dfrac{p_e}{p*} = 1.94 \times \dfrac{1}{5.48} \times \dfrac{1}{0.094} \times \dfrac{272}{660} = 1.552$

From Fanno tables Approximately

At $\quad \dfrac{p_e}{p^*} = 1.552, M_e = 0.68, \dfrac{\lambda_{av} L_{max}}{D} = 0.25$

We know the relation

$$\frac{\lambda_{av} L}{D} = \left(\frac{\lambda_{av} L_{max}}{D}\right)_{My} - \left(\frac{\lambda_{av} L_{max}}{D}\right)_{Me}$$

$$= 0.747 - 0.25 = 0.497$$

Also $\dfrac{\dot{m}}{A} = \dfrac{p}{\sqrt{T_o}} \times \sqrt{\dfrac{k}{R}} \, M \sqrt{1 + \dfrac{k-1}{2} M^2}$

$$= \frac{62.04 \times 1000}{\sqrt{325}} \times \sqrt{\frac{1.4}{287}} \times 2.2 \sqrt{1 + 0.2\,(2.2)^2} = 741.8 \; kg/s.$$

(b) From Fanno tables

At $\quad M_2 = 2.2, \dfrac{\lambda_{av} L_{max}}{D} = 0.361, \dfrac{p_2}{p^*} = 0.355$

Also, $\dfrac{p_e}{p^*} = \dfrac{p_e}{p_o} \times \dfrac{p_o}{p_2} \times \dfrac{p_2}{p^*}$

$$= \frac{272}{660} \times \frac{1}{0.094} \times 0.355 = 1.5564$$

From Fanno tables

At $\quad \dfrac{p_e}{p^*} = 1.5564, M_{ey} = 0.68$

From normal shock tables

At $\quad M_{ey} = 0.68\,, M_{ex} = 1.56$

From Fanno tables

At $\quad M_{ex} = 1.56, \dfrac{\lambda_{av} L_{\max}}{D} = 0.158$

We know that

$$\therefore \frac{\lambda_{av} L}{D} = \left(\frac{\lambda_{av} L_{\max}}{D}\right)_{M_2} - \left(\frac{\lambda_{av} L_{\max}}{D}\right)_{M_{ex}}$$

$$= 0.361 - 0.158 = 0.203$$

Also $\dfrac{\dot{m}}{A} = \dfrac{p}{\sqrt{T_o}} \times \sqrt{\dfrac{k}{R}}\, M \sqrt{1 + \dfrac{k-1}{2} M^2} = 741.8 \text{ kg/s}$ Ans

6.33 An insulated pipe of length L and diameter D discharges into region where the static pressure is 4 bar. Air flows from a convergent-divergent nozzle with a throat area half of that of pipe. The stagnation pressure at the nozzle inlet is 10 bar. The friction factor for the pipe is 0.02 and L/D = 20. If shock occurs in between the pipe, find the Mach number at exit from the pipe.

Solution

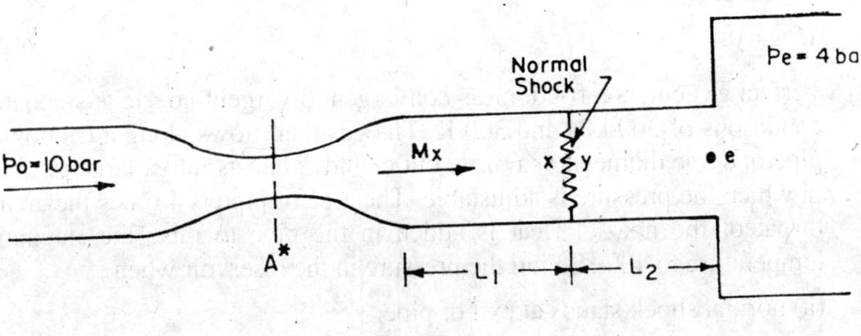

Fig. 6.29 Convergent - divergent nozzle with insulated pipe P 6.33.

From isentropic tables

At $\quad \dfrac{A}{A^*} = 2,\ M_i = 2.2,\ \dfrac{p_i}{p_o} = 0.094$

Therefore, $p_i = 0.094 \times 10 = 0.94\ bar$

Intensity of shock depends on the ratio of static pressure at exit and inlet of the pipe.

Therefore, $\frac{p_e}{p} = \frac{p_y}{p_x} = \frac{4}{0.94} = 4.2553$

Approximately

At $\frac{p_y}{p_x} = 4.2,\ M_x = 1.94,$

$M_x = 0.5880$

Also, We have

$$\frac{\lambda_{av} L}{D} = \left(\frac{\lambda_{av} L_{max}}{D}\right)_{Mi} - \left(\frac{\lambda_{av} L_{max}}{D}\right)_{Mx}$$

$$+ \left(\frac{\lambda_{av} L_{max}}{D}\right)_{My} - \left(\frac{\lambda_{av} L_{max}}{D}\right)_{Me}$$

From Fanno tables

$$0.02 \times 20 = 0.361 - 0.287 + 0.5335 - \left(\frac{\lambda_{av} L_{max}}{D}\right)_{Me}$$

$$\left(\frac{\lambda_{av} L_{max}}{D}\right)_{Me} = 0.361 - 0.287 + 0.5335 - 0.4 = 0.2075$$

From Fanno tables

$M_e = 0.7$ Ans

6.34 A perfect gas enters a frictionless convergent divergent nozzle at stagnation conditions of 300 kPa and 300 K. The gas then flows along a frictionless pipe of some diameter as nozzle outlet and exhausts into a large reservoir in which the pressure is adjustable. The area of pipe is 4 times the area of throat of the nozzle. Heat is added in the pipe to raise the stagnation temperature to 400 K. Find the pressure in the reservoir when

(a) normal shock stands at exit of pipe.

(b) Weak normal shock is formed just downstream of the throat.

Solution

For the nozzle section, from isentropic tables

At $\frac{A_1}{A^*} = 4,\ M_1 = 2.94,\ \frac{p_1}{p_o} = 0.0298,\ \frac{T_1}{T_o} = 0.366$

Therefore, $p_1 = 300 \times 0.0298 = 8.94\ kPa$

$T_1 = 300 \times 0.366 = 110\ K$

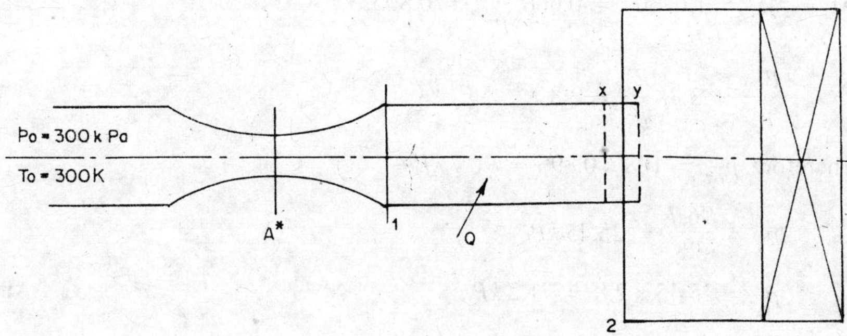

Fig. 6.30 Frictionless convergent nozzle P 6.34.

Further the pipe is heated and therefore from Rayleigh tables

At $M_1 = 2.94, \left(\frac{T_o}{T_o^*}\right)_1 = 0.66$

and $\left(\frac{p_o}{p_o^*}\right)_1 = 3.251$

Also, $\left(\frac{T_o}{T_o^*}\right)_2 = \left(\frac{T_o}{T_o^*}\right)_1 \times \frac{T_{o2}}{T_{o1}} = 0.660 \times \frac{400}{300} = 0.88$

From Rayleigh tables

At $\left(\frac{T_o}{T_o^*}\right)_2 = 0.88,\ M_2 = 1.6$

and $\left(\frac{p_o}{p_o^*}\right)_2 = \left(\frac{p_o}{p_o^*}\right)_1 \times \frac{p_{o2}}{p_{o1}}$

Therefore, $\frac{p_{o2}}{p_{o1}} = \frac{1.176}{3.251} = 0.361$

$$p_{o2} = 0.361 \times 300 = 108\ kPa$$

Also, $p_{o2} = p_{ox} = 108\ kPa$

From normal shock tables,

At $M_x = 1.6,\ M_y = 0.668,\ \dfrac{p_{oy}}{p_{ox}} = 0.895,$

$$\frac{p_{oy}}{p_x} = 3.8,\ \frac{p_y}{p_x} = 2.8$$

Therefore, $p_{oy} = 108 \times 0.895 = 96.7\ kPa$

$$p_x = \frac{96.7}{3.8} = 25.45\ kPa$$

$$p_y = 25.45 \times 2.8 = 71.2\ kPa$$ Ans

(b) Weak shock is an isentropic one

From isentropic tables

$$\frac{A}{A^*} = 4,\ M_1 = 0.15,\ \frac{p_1}{p_o} = 9.84$$

$$p_1 = 0.984 \times 300 = 295.2\ kPa$$

This is pressure available if a weak shock is considered isentropic

From Rayleigh tables

At $M_1 = 0.15, \left(\dfrac{T_o}{T_o^*}\right)_1 = 0.0978, \left(\dfrac{p}{p^*}\right)_1 = 2.3265$

Also, $\dfrac{T_{o2}}{T_{o1}} = \dfrac{400}{300} = \dfrac{(T_o/T_o^*)_2}{(T_o/T_o^*)_1}$

$$\left(\frac{T_o}{T_o^*}\right)_2 = \frac{400}{300} \times 0.0978 = 0.1304$$

From Rayleigh tables

At $\left(\dfrac{T_o}{T_o^*}\right)_2 = 0.1304,\ M_2 = 0.17, \left(\dfrac{p}{p^*}\right)_2 = 2.307$

$$\frac{p_2}{p_1} = \frac{(p/p^*)_2}{(p/p^*)_1} = \frac{2.307}{2.3265} = 0.9916$$

Therefore, $p_2 = 0.9916 \times 295.2 = 292.72\ kPa$ Ans

EXERCISES

A. THEORY

6.1 Define normal shock.

6.2 What assumptions are used for deriving the normal shock governing equation? Derive the governing equation of normal shock.

6.3 Draw T–S or h–s chart showing the relationship between a normal shock and Fanno and Rayleigh lines.

6.4 Explain why the normal discontinuity always involves a change from supersonic to subsonic speed with a consequent pressure rise, and never the reverse ?

6.5 Explain how strong compression and expansion waves are formed in a compressible fluid.

6.6 Under what conditions a compression wave changes into a shock wave ?

6.7 Why are expansion shocks impossible ?

6.8 Write down the four basic equations which satisfy the state points before and after a normal shock wave.

6.9 Prove that for normal shock at the maximum entropy point on the h–s or T–S diagrams the Mach number is unity.

6.10 Derive the Prandtl-Meyer relation

$$V_x V_y = a^{*2} \quad or \quad M_x^* M_y^* = 1$$

6.11 Derive the following relations for normal shock waves

(a) $$(M_y)_{minimum} = \sqrt{\frac{k+1}{2k}}$$

(b) $$\left(\frac{\rho_y}{\rho_x}\right)_{maximum} = \frac{k+1}{k-1}$$

6.12 Derive the following relations for flow through a normal shock in a perfect gas.

(a) $$\frac{p_y}{p_x} = \frac{1 + k M_x^2}{1 + k M_y^2}$$

(b) $$M_y^2 = \frac{2/k - 1 + M_x^2}{\frac{2k}{k-1} M_x^2 - 1}$$

(c) $$\frac{p_y}{p_x} = \frac{2k}{k+1} M_x^2 - \frac{k-1}{k+1}$$

(d) $$\frac{T_y}{T_x} = \frac{\left(1 + \frac{k-1}{2} M_x^2\right)\left(\frac{2k}{k-1} M_x^2 - 1\right)}{\frac{(k+1)^2}{2\,(k-1)} M_x^2}$$

(e) $$\frac{\rho_y}{\rho_x} = \frac{V_x}{V_y} = \frac{(k+1)\, M_x^2}{2 + (k-1)\, M_x^2}$$

(f) $$\frac{V_y}{V_x} = \left(\frac{2}{k+1}\right)\frac{1}{M_y^2} + \frac{k-1}{k+1}$$

6.13 Derive the Rankine Hugoniot relation

$$\frac{p_y}{p_x} = \frac{\frac{\rho_y}{\rho_x}\left(\frac{k+1}{k-1}\right) - 1}{\left(\frac{k+1}{k-1}\right) - \frac{\rho_y}{\rho_x}}$$

Draw the graph between ρ_y/ρ_x and p_y/p_x. From the graph nearly isentropic.

6.14 Depict graphically the variation of the following parameters with the upstream Mach number of the normal shock : M_y, p_y/p_x, p_{oy}/p_{ox}, $\frac{T_y}{T_x}$, V_x/V_y.

6.15 Discuss the change across normal shock wave characteristics of the following parameters : M, p, p_o, T, T_o ρ and V.

6.16 Prove the formula

(a) $$\frac{p_{oy}}{p_{ox}} = e^{-\Delta S/R}$$

(b) $$\frac{s_y - s_x}{R} = \frac{k}{k-1} \ln\left[\frac{2}{(k+1)} M_x^2 + \frac{k-1}{k+1}\right] + \frac{1}{k-1} \ln\left[\frac{2k}{k+1} M_x^2 - \frac{k-1}{k+1}\right]$$

6.17 The entropy change is always positive when M_x is greater than unity and is always negative when M_x is less than unity. Explain.

6.18 Explain why for a perfect gas only the shock from supersonic to subsonic speed is possible ?

6.19 Explain why the shock waves cannot develop in a subsonic flow.

6.20 Define the following : Weak shock waves, strength of shock wave shock of vanishing strength and irreversibility of weak shock.

6.21 Prove that the strength of the shock wave

$$p = \frac{2k}{k+1}\left(M_x^2 - 1\right) = \frac{\left(\frac{2k}{k-1}\right)\left(\frac{\rho_y}{\rho_x} - 1\right)}{\frac{2}{k-1} - \left(\frac{\rho_y}{\rho_x} - 1\right)}$$

6.22 Prove that for weak shock, the strength of the shock is proportional to $\left(\frac{\rho_y}{\rho_x} - 1\right)$

$$p \propto \left(\frac{\rho_y}{\rho_x} - 1\right)$$

6.23 Derive the Rayleigh supersonic pitot tube formula

$$\frac{p_{oy}}{p_{ox}} = \left(\frac{k+1}{2} M_x^2\right)^{k/k-1} \left(\frac{2k}{k+1} M_x^2 - \frac{k-1}{k+1}\right)^{1/k-1}$$

6.24 Define diffuser efficiency. What are the assumptions used for defining the diffuser efficiency ?

6.25 Prove that the diffuser efficiency is only a function of the entry Mach number.

6.26 Explain why the throat area of the diffuser in a supersonic wind tunnel is larger than the nozzle throat area. Why is it desirable to locate the shock wave at the diffuser throat during the operation of the tunnel ?

6.27 What are the two difficulty arises when it is assumed that the supersonic diffuser is a reverse of convergent divergent nozzle.

6.28 Discuss the most favourable operating condition and best operating condition for starting of supersonic wind tunnel diffuser.

6.29 Draw a curve between area contraction ratio V/S test section Mach number M_x and discuss.

6.30 Draw a curve between M_1 (Inlet Mach number) V/S diffuser efficiency showing best fixed geometry diffuser and best normal shock diffuser and discuss.

6.31 Explain why the pressure of normal shocks in a converging-diverging nozzle from a reservoir does not affect the flow rate through the nozzle.

6.32 A converging-diverging nozzle is attached to a pressure tank containing a gas at stagnation conditions. Explain why a shock cannot occur in a cnverging portion of the nozzle. Assume isentropic flow, except across the shock.

B. UNSOLVED PROBLEMS

6.1 Use table to obtain answers to the following situations involving normal shock in air.

(a) The stagnation pressure on the noss of a blunt nosed projectile is 3.5 times the free stream pressure of the atmosphere. What is the free flight Mach number ?

(b) What is the Mach number just downstream of the shock in part (a) ?

(c) The stagnation of a shock may be defined as the ratio of the pressure increase across it to the upstream pressure $\left(p_y - p_x\right)/p_x$. What Mach number would produce a shock of strength 5 ?

(d) The ratio of downstream to upstream stagnation pressure across a normal shock is a direct measure of its irreversibility. The entropy increase is $\Delta s = -R \ln\left(p_{oy}/p_{ox}\right)$. At what Mach number is the entropy increase 218 J/kg K ? [(a) 1.523, (b) 0.692, (c) 2.299, (d) 2.58]

6.2 Use tables in solving the following situations involving air flow. Consider the flow to be isentropic except across shocks.

(a) A normal shock occurs in a converging-diverging nozzle fed from a reservoir. The shock occurs when M = 2.2. What is the ratio of the pressure just downstream from the shock to the reservoir pressure p_y/p_{ox}?

(b) What is the stagnation temperature just downstream from the shock in terms of the reservoir temperature ?

(c) What is the area of the nozzle where the shock forms in part (a) in terms of the throat area ?

(d) What is the exit area for conditions of part (a) in terms of the throat area for an exit pressure 60 percent of the reservoir pressure ($p_e = 0.6\ p_o$)

[(a) 0.5125, (b) $T_{ox} = T_{oy}$, (c) M = 2.2, $\frac{A_x}{A_x^*} = 2.005$, (d) $\frac{A_e}{A_x^*} = 3.72$]

6.3 (a) Does the pressure ratio across a normal shock p_y/p_x increase or decrease as the upstream Mach number M_x increase ?

(b) Given the pressure p_x, the temperature T_x, the Mach number M_x, and ratio of specific heat capacities k for a gas, which the these must be knwon in order that the Mach number My downstream of a normal shock may be calculated ?

(c) At a Mach number of 0.8, is the stagnation pressure for a given flow greater, equal, or less if compressible effects are considered than if they ignored ?

(d) The Mach number at a given location in a diverging duct is 1.4. Does the Mach numebr increase, remain the same, or decrease downstream ? No shocks are present.

(e) Repeat part (d) for a Mach number of 0.8

[(a) Increase, (b) M_X and K, (c) Greater, (d) Increase, (e) Decrease.]

6.4 A normal shock stands in a duct. The fluid is air, which can be considered an ideal gas properties upstream from the shock are $t_1 = 5^oC$ $p_1 = 65$ kPa, $V_1 = 668$ m/s. Determine the properties downstream and $s_2 - s_1$. 0.0939 kJ/kg K. [469 k, 293 kPa, 2.18 kg/m^3, 251 m/s, 500 K, 367 kPa].

6.5 An air stream with temperature 273 k, pressure 60 kPa and speed 497 m/s, undergoes a normal shock. The temperature downstream from the shock is 87°C. Determine the Mach number, speed, and stagnation pressure downstream from the shock. [$M_y = 0.707$, $V_y = 269$ m/s, $p_{oy} = 205$ kPa]

6.6 Show that the downstream Mach number for a normal shock approach a limiting minimum value as the upstream Mach number increases without limit. What is the minimum Mach number for (a) Air ? (b) Carbon oxide ? (c) Xenon ? [$My_{min} = 0.378$ for air, 0.34 for Co_2, 0.446 for Xenon]

6.7 A total pressure tube used to measure the air speed in a supersonic wind tunnel measures a pressure seven times the static pressure. What is the Mach number of the flow. [$M_x = 2.25$].

6.8 The density of air is quardrupled in passing through a normal shock. What is the Mach number of the upstream flow ? [$M_x = 3.18$]

6.9 An air stream at 4°C static temperature has a stagnation temperature of 490 K. What is the Mach number just downstream of a normal shock ? [$T/T_o = 0.565$; $M_x = 1.968$ and $M_y = 0.583$]

6.10 A normal shock occurs in air at a section where the flow speed is 924 m/s. The temperature and pressure at this point are 283 K and 35 kPa. The absolute pressure downstream from the shock is 301 kPa. Determine the speed and Mach number downstream from the shock. [$V_y = 256$ m/s, $M_y = 0.493$]

6.11 Air approaches a normal shock at 273 K, 60 kPa and 497 m/s. The air speed immediately downstream from the shock is 267 m/s. Evaluate all flow properties downstream from the shock and the entropy change across the shock. [$T_y = 360.5$ K, $M_y = 0.702$, $\rho_y = 1.426$ kg/m^3 and $p_{o2} = 205$ kPa, $T_{ox} = T_{oy} = 396$ k $p_2 = 147.5$ kPa]

6.12 An air stream approaches a normal shock at Mach number, $M_x = 2.64$. The upstream stagnation pressure and density are $p_{ox} = 3$ MPa and $\rho_x = 1.65$ kg/m^3, respectively. The ratio of static pressure to stagnation pressure immediately behind the shock is 0.843. Determine the downstream Mach number and temperature. [$M_y = 0.5$, $T_y = 679$ K].

6.13 A supersonic aircraft flies at Mach number 2.7 at 20 km altitude on a standard where it is slowed isentropically to Mach number 1.3. A normal shock occurs at that location. (The Mach number immediately downstream from the shock is 0.786). The resulting subsonic flow is decelerated further to a Mach number of 0.4. The subsonic diffusion is adiabatic but not isentropic, the final pressure is 104 kPa. Evaluate (a) the stagnation temperature for the flow, (b) the pressure change across the shcok, (c) the entropy change, and (d) the final stagnation pressure. [(a) $T_{ox} = T_{oy} = 533$ K, (b) $\Delta p = 37.4$ kPa, (c) 30 J/kg K, (d) 116 kPa.]

6.14 A supersonic aircraft flies at Mach number 2.7 at an altitude of 20 km on a standard day. Air approaching a forward facing total head (pitot) tube passes through a normal shock and then is brought isentropically to rest relative to the aircraft. Determine the final temperature of the air and the pressure sensed by the total head tube. [$T_{oy} = T_{ox} = 533$ K, 54.6 kPa]

6.15 A supersonic aircraft flies at an altitude of 22 km on a standard day. A forward facing stagnation tube senses a total pressure of 53.3 kPa. Determine the Mach number and speed of the aircraft. [M = 3.2, V = 949 m/s].

6.16 A supersonic aircraft cruises at Mach number at 12 km altitude. A pitot tube is used to sense stagnation pressure for calculating air speed. A normal shock stands in front of the pitot tube; the Mach number following the shock is 0.5471. Evaluate local isentropic stagnation conditions in front of the shock. Estimate the pressure sensed by the pitot tube. [426.5 K, 207 kPa, 106.3 kPa].

6.17 A convergent-divergent nozzle is attached to a large tank of air in which the temperature and pressure are 300 K and 250 kPa. At the nozzle throat the pressure is 132 kPa. In the diverging section the pressure falls to 68.1 kPa before rising suddenly across a normal shock. At the nozzle exit the pressure is 180 kPa. Find the Mach number immediately behind the shock. Determine the pressure immediately downstream from the shock. Calculate the entropy change across the shock. [$M_y = 0.7011$, $p_y = 167.4$ kPa, 20.9 J/kg K].

6.18 A convergent divergent nozzle with air ratio, Ae/At = 1.633, is designed to operate with atmospheric pressure at the exit plane. Determine the range(s) of stagnation pressures for which the stagnation pressure for which the nozzle will before from normal shocks. [(b) $M_e = 1.96$, $p_o >$ 743 kPa (a) $O < M_e < 0.388$, $p_{atm} < p_o < 112$ kPa]

6.19 Air discharges from a reservoir into atmosphere through a convergent divergent nozzle of circular cross section. Compute :

(a) the diameter of the nozzle at exit for shockless flow condition if M = 2 is to be attained, assuming isentropic flow and throat diameter 12.5 mm.

(b) the correct pressure in the reservoir assuming the atmospheric pressure to be 1.01325 bar. The temperature of air in the reservoir is 21°C.

(c) If the jet impings on a flat plate placed vertically to the flow, calculate the force of the impact. [16.5 mm, 0.132 bar, 123.7 N]

6.20 A supersonic nozzle is provided with constant diameter, circular duct at its exit. The duct diameter is same as the nozzle exit diameter. Nozzle exit cross section is three times that of its throat. The entry conditions of the gas. (k = 1.4, R = 0.287 T kJ/kg K) are p_o = 10 bar, T_o = 600 K. Calculate the static pressure, Mach number and the velocity of the gas in the duct :

(a) When the nozzle operates at its design condition,

(b) when a normal shock occurs at its exit, and

(c) when a normal shock occurs at a section in the diverging, part where the area ratio $A/A^* = 2$.

(a) $p = 0.471\ bar,\ M = 2.64,\ V = 837.05\ m/s$

(b) $p = 3.7515\ bar,\ M = 0.5,\ V = 239.33\ m/s$

(c) $p = 5.834\ bar,\ M = 0.328,\ V = 159.3\ m/s$

6.21 A Mach–2 aircraft engine employes a subsonic inlet diffuser of area ratio 3. A normal shock is formed just upstream of the diffuser inlet. The free stream conditions upstream of the diffuser are : p = 0.1 bar, T = 300 K. Determine

(a) Mach number, pressure and temperature at the diffuser exit.

(b) Diffuser efficiency including the shcok.

Assume isentropic flow in the diffuser downstream of the shock.

[(a) M = 0.16, p = 0.554 bar, T = 537.8 K (b) η_D = 80 %]

6.22 Air is drawn from a standard atmosphere through a converging diverging nozzle attached to a vacuum tank. The throat area is 10 cm^2, and the exit area at the vacuum tank is 16 cm^2. A normal shock occurs where the nozzle area is 13 cm^2 what is the pressure in the tank ?

[M_x = 1.659, M_y = 0.6515, p_{tank} = 0.872 atmosphere]

6.23 Air flow from a reservoir at 600 kPa and 50°C through a converging-diverging nozzle. A shock occurs at a section where the pressure is 150 kPa.

(a) What is the static pressure just beyond the shock ?

(b) What is the stagnation pressure downstream of the shock ? Note the decrease in stagnation pressure.

[(a) p_y = 400 kPa, p_{oy} = 546 kPa]

Table-1 One-Dimensional Isentropic Compressible-Flow Functions

For a perfect gas with constant specific heat and molecular weight k = 1.4

M	M^*	$\frac{A}{A^*}$	$\frac{p}{p_o}$	$\frac{\rho}{\rho_o}$	$\frac{T}{T_o}$	$\frac{F}{F^*}$	$\left(\frac{A}{A^*}\right)\left(\frac{p}{p_o}\right)$
o	0	∞	1.00000	1.00000	1.00000	∞	∞
0.01	0.01096	57.874	.99993	.99995	.99998	45.650	57.870
.02	.02191	28.942	.99972	.99980	.99992	22.834	28.934
.03	.03286	19.300	.99937	.99955	.99982	15.232	19.288
.04	.04381	14.482	.99888	.99920	.99968	11.435	14.465
.05	.05476	11.592	.99825	.99875	.99950	9.1584	11.571
.06	.06570	9.6659	.99748	.99820	.99928	7.6428	9.6415
.07	.07664	8.2915	.99658	.99755	.99902	6.5620	8.2631
.08	.08758	7.2616	.99553	.99680	.99872	5.7529	7.2291
.09	.09851	6.4613	.99435	.99596	.99838	5.1249	6.4248
.10	.10943	5.8218	.99303	.99502	.99800	4.6236	5.7812
.11	.12035	5.2992	.99157	.99398	.99758	4.2146	5.2546
.12	.13126	4.8643	.98998	.99284	.99714	3.8747	4.8157
.13	.14216	4.4968	.98826	.99160	.99664	3.5880	4.4440
.14	.15306	4.1824	.98640	.99027	.99610	3.3432	4.1255
.15	.16395	3.9103	.98441	.98884	.99552	3.1317	3.8493
.16	.17483	3.6727	.98228	.98731	.99490	2.9474	3.6076
.17	.18569	3.4635	.98003	.98569	.99425	2.7855	3.3943
.18	.19654	3.2779	.97765	.98398	.99356	2.6422	3.2046
.19	.20738	3.1122	.97514	.98217	.99283	2.5146	3.0348
.20	.21822	2.9635	.97250	.98027	.99206	2.4004	2.8820
.21	.22904	2.8293	.96973	.97828	.99125	2.2976	2.7437
.22	.23984	2.7076	.96685	.97621	.99041	2.2046	2.6178
.23	.25063	2.5968	.96383	.97403	.98953	2.1203	2.5029
.24	.26141	2.4956	.96070	.97177	.98861	2.0434	2.3975
.25	.27216	2.4027	.95745	.96942	.98765	1.9732	2.3005
.26	.28291	2.3173	.95408	.96699	.98666	1.9088	2.2109
.27	.29364	2.2385	.95060	.96446	.98563	1.8496	2.1279
.28	.30435	2.1656	.94700	.96185	.98456	1.7950	2.0508
.29	.31504	2.0979	.94329	.95916	.98346	1.7446	1.9789

M	M^*	$\frac{A}{A^*}$	$\frac{p}{p_o}$	$\frac{\rho}{\rho_o}$	$\frac{T}{T_o}$	$\frac{F}{F^*}$	$\left(\frac{A}{A^*}\right)\left(\frac{p}{p_o}\right)$
.30	.32572	2.0351	.93947	.95638	.98232	1.6979	1.9119
.31	.33638	1.9765	.93554	.95352	.98114	1.6546	1.8491
.32	.34701	1.9218	.93150	.95058	.97993	1.6144	1.7902
.33	.35762	1.8707	.92736	.94756	.97868	1.5769	1.7348
.34	.36821	1.8229	.92312	.94446	.97740	15420.	1.6828
.35	.37879	1.7780	.91877	.94128	.97608	1.5094	1.6336
.36	.38935	1.7358	.91433	.93803	.97473	1.4789	1.5871
.37	.39988	1.6961	.90979	.93470	.97335	1.4503	1.5431
.38	.41039	1.6587	.90516	.93129	.97193	1.4236	1.5014
.39	.42087	1.6234	.90044	.92782	.97048	1.3985	1.4618
0.40	.43133	1.5901	.89562	.92428	.96899	1.3749	1.4241
.41	.44177	1.5587	.89071	.92066	.96747	1.3527	1.3883
.42	.45218	1.5289	.88572	.91697	.96592	1.3318	1.3542
.43	.46256	1.5007	.88065	.91322	.96434	1.3122	1.3216
.44	.47292	1.4740	.87550	.90940	.94272	1.2937	1.2905
.45	.48326	1.4487	.87027	.90552	.96108	1.2763	1.2607
.46	.49357	1.4246	.86496	.90157	.95940	1.2598	1.2322
.47	.50385	1.4018	.85958	.89756	.95769	1.2443	1.2050
.48	.51410	1.3801	.85413	.89349	.95595	1.2296	1.1788
.49	.52432	1.3594	.84861	.88936	.95418	1.2158	1.1537
.50	.53452	1.3398	.84302	.88517	.95238	1.2027	1.12951
.51	.54469	1.3212	.83737	.88092	.95055	1.1903	1.10631
.52	.55482	1.3034	.83166	.87662	.94869	1.1786	1.08397
.53	.56493	1.2864	.82589	.87227	.94681	1.1675	1.06245
.54	.57501	1.2703	.82005	.86788	.94489	1.1571	1.04173
.55	.58506	1.2550	.81416	.86342	.94295	1.1472	1.02174
.56	.59508	1.2403	.80822	.85892	.94098	1.1378	1.00244
.57	.60506	1.2263	.80224	.85437	.93898	1.1289	.98381
.58	.61500	1.2130	.79621	.84977	.93696	1.1205	.96581
.59	.62491	1.2003	.79012	.84513	.93491	1.1126	.94839

M	M^*	$\frac{A}{A^*}$	$\frac{p}{p_o}$	$\frac{\rho}{\rho_o}$	$\frac{T}{T_o}$	$\frac{F}{F^*}$	$\left(\frac{A}{A^*}\right)\left(\frac{p}{p_o}\right)$
.60	.63480	1.1882	.78400	.84045	.93284	1.10504	.93155
.61	.64466	1.1766	.77784	.83573	.93074	1.09793	.91525
.62	.65448	1.1656	.77164	.83096	.92861	1.09120	.89946
.63	.66427	1.1551	.76540	.82616	.92616	1.08485	.88416
.64	.67402	1.1451	.75913	.82132	.92428	1.07883	.86932
.65	.68374	1.1356	.75283	.81644	.92208	1.07314	.85493
.66	.69342	1.1265	.74650	.81153	.91986	1.06777	.84096
.67	.70307	1.1178	.74014	.80659	.91762	1.06271	.82740
.68	.71268	1.1096	.73376	.80162	.91535	1.05792	.81421
.69	.72225	1.1018	.72735	.79662	.91306	1.05340	.80141
.70	.73179	1.09437	.72092	.79158	.91075	1.04915	.78896
.71	.74129	1.08729	.71448	.78652	.90842	1.04514	.77685
.72	.75076	1.08057	.70802	.78143	.90606	.104137	.76507
.73	.76019	1.07419	.70155	.77632	.90368	1.03783	.75360
.74	.76958	1.06814	.69507	.77119	.90129	1.03450	.74243
.75	.77893	1.06242	.68857	.76603	.89888	1.03137	.73155
.76	.78825	1.05700	.68207	.76086	.89644	1.02844	.72095
.77	.79753	1.05188	.67556	.75567	.89399	1.02570	.71062
.78	.80677	1.04705	.66905	.75046	.89152	1.02314	.70054
.79	.81597	1.04250	.66254	.74524	.88903	1.02075	.69070
0.80	.82514	1.03823	.65602	.74000	.88652	1.01853	.68110
.81	.83426	1.03422	.64951	.73474	.88400	1.01646	.67173
.82	.84334	1.03046	.64300	.72947	.88146	1.01455	.66259
.83	.85239	1.02696	.63650	.72419	.87890	1.01278	.65366
.84	.86140	1.02370	.63000	.71890	.87633	1.01115	.64493
.85	.87037	1.02067	.62351	.71361	.87374	1.00966	.63640
.86	.87929	1.01787	.61703	.70831	.87114	1.00829	.62806
.87	.88817	1.01530	.61057	.70300	.86852	1.00704	.61991
.88	.89702	1.01294	.60412	.69769	.86589	1.00591	.61193
.89	.90583	1.01080	.59768	.69237	.86324	1.00490	.60413

M	M^*	$\frac{A}{A^*}$	$\frac{p}{p_o}$	$\frac{\rho}{\rho_o}$	$\frac{T}{T_o}$	$\frac{F}{F^*}$	$\left(\frac{A}{A^*}\right)\left(\frac{p}{p_o}\right)$
.90	.91460	1.00886	.59126	.68704	.86058	1.00399	.59650
.91	.92333	1.00713	.58486	.68171	.85791	1.00318	.58903
.92	.93201	1.00560	.57848	.67639	.85523	1.00248	.58171
.93	.94065	1.00426	.57212	.67107	.85253	1.00188	.57455
.94	.94925	1.00311	.56578	.66575	.84982	1.00136	.56754
.95	.95781	1.00214	.55946	.66044	.84710	1.00093	.56066
.96	.96633	1.00136	.55317	.65513	.84437	1.00059	.55392
.97	.97481	1.00076	.54691	.64982	.84162	1.00033	.54732
.98	.98325	1.00033	.54067	.64452	.83887	1.00014	.54085
.99	.99165	1.00008	.53446	.63923	.83611	1.00003	.53450
1.00	1.00000	1.00000	.52828	.63394	.83333	1.00000	.52828
1.01	1.00831	1.00008	.52213	.62866	.83055	1.00003	.52218
1.02	1.01658	1.00033	.51602	.62339	.82776	1.00013	.51619
1.03	1.02481	1.00074	.50994	.61813	.82496	1.00030	.51031
1.04	1.03300	1.00130	.50389	.61288	.82215	1.00053	.50454
1.05	1.04114	1.00202	.49787	.60765	.81933	1.00082	.49888
1.06	1.04924	1.00290	.49189	.60243	.81651	1.00116	.49332
1.07	1.05730	1.00394	.48595	.59722	.81368	1.00155	.48787
1.08	1.06532	1.00512	.48005	.59203	.81084	1.00200	.48251
1.09	1.07330	1.00645	.47418	.58685	.80800	1.00250	.47724
1.10	1.08124	1.00793	.46835	.58169	.80515	1.00305	.47206
1.11	1.08914	1.00955	.46256	.57655	.80230	1.00365	.46698
1.12	1.09699	1.01131	.45682	.57143	.79944	1.00429	.46199
1.13	1.10480	1.01322	.45112	.56632	.79657	1.00497	.45708
1.14	1.11256	1.01527	.44545	.56123	.79370	1.00569	.45225
1.15	1.1203	1.01746	.43983	.55616	.79083	1.00646	.44751
1.16	1.1280	1.01978	.43425	.55112	.78795	1.00726	.44284
1.17	1.1356	1.02224	.42872	.54609	.78507	1.00810	.43825
1.18	1.1432	1.02484	.42323	.54108	.78218	1.00897	.43374
1.19	1.1508	1.02757	.41778	.53610	.77929	1.00988	.42930

M	M^*	$\frac{A}{A^*}$	$\frac{p}{p_o}$	$\frac{\rho}{\rho_o}$	$\frac{T}{T_o}$	$\frac{F}{F^*}$	$\left(\frac{A}{A^*}\right)\left(\frac{p}{p_o}\right)$
1.20	1.1583	1.03044	.41238	.53114	.77640	1.01082	.42493
1.21	1.1658	1.03344	.40702	.52620	.77350	1.01178	.42063
1.22	1.1732	1.03657	.40171	.52129	.77061	1.01278	.41640
1.23	1.1806	1.03983	.39645	.51640	.76771	1.01381	.41224
1.24	1.1879	1.04323	.39123	.51154	.76481	1.01486	.40814
1.25	1.1952	1.04676	.38606	.50670	.76190	1.01594	.40411
1.26	1.2025	1.05041	.38094	.50189	.75900	1.01705	.40014
1.27	1.2097	1.05419	.37586	.49710	.75610	1.01818	.39622
1.28	1.2169	1.05810	.37083	.49234	.75319	1.01933	.39237
1.29	1.2240	1.06214	.36585	.48761	.75029	1.02050	.38858
1.30	1.2311	1.06631	.36092	.48291	.74738	1.02170	.38484
1.31	1.2382	1.07060	.35603	.47823	.74448	1.02292	.38116
1.32	1.2452	1.07502	.35119	.47358	.74158	1.02415	.37754
1.33	1.2522	1.07957	.34640	.46895	.73867	1.02540	.37397
1.34	1.2591	1.08424	.34166	.46436	.73577	1.02666	.37044
1.35	1.2660	1.08904	.33697	.45980	.73287	1.02794	.36697
1.36	1.2729	1.09397	.33233	.45527	.72997	1.02924	.36355
1.37	1.2797	1.09902	.32774	.45076	.72707	1.03056	.36018
1.38	1.2865	1.10420	.32319	.44628	.72418	1.03189	.35686
1.39	1.2932	1.10950	.31869	.44183	.72128	1.03323	.35359
1.40	1.2999	1.1149	.31424	.43742	.71839	1.03458	.35036
1.41	1.3065	1.1205	.30984	.43304	.71550	1.03595	.34717
1.42	1.3131	1.1262	.30549	.42869	.71261	1.03733	.34403
1.43	1.3197	1.1320	.30119	.42436	.70973	1.03872	.34093
1.44	1.3262	1.1379	.29693	.42007	.70685	1.04012	.33787
1.45	1.3327	1.1440	.29272	.41581	.70397	1.04153	.33486
1.46	1.3392	1.1502	.28856	.41158	.70110	1.04295	.33189
1.47	1.3456	1.1562	.28445	.40738	.69823	1.04438	.32896
1.48	1.3520	1.1629	.28039	.40322	.69537	1.04581	.32607
1.49	1.3583	1.1695	.27637	.39909	.69251	1.04725	.32321

M	M^*	$\frac{A}{A^*}$	$\frac{p}{p_o}$	$\frac{\rho}{\rho_o}$	$\frac{T}{T_o}$	$\frac{F}{F^*}$	$\left(\frac{A}{A^*}\right)\left(\frac{p}{p_o}\right)$
1.50	1.3646	1.1762	.27240	.39498	.68965	1.04870	.32039
1.51	1.3708	1.1830	.26848	.39091	.68680	1.05016	.31761
1.52	1.3770	1.1899	.26461	.38687	.68396	1.05162	.31487
1.53	1.3832	1.1970	.26078	.38287	.68112	1.05309	.31216
1.54	1.3894	1.2042	.25700	.37890	.67828	1.05456	.30948
1.55	1.3955	1.2115	.25326	.37496	.67545	1.05604	.30685
1.56	1.4016	1.2190	.24957	.37105	.67262	1.05752	.30424
1.57	1.4076	1.2266	.24593	.36717	.66980	1.05900	.30167
1.58	1.4135	1.2343	.24233	.36332	.66699	1.06049	.29913
1.59	1.4195	1.2422	.23878	.35951	.66418	1.06198	.29662
1.60	1.4254	1.2502	.23527	.35573	.66138	1.06348	.29414
1.61	1.4313	1.2583	.23181	.35198	.65858	1.06498	.29169
1.62	1.4371	1.2666	.22839	.34826	.65579	1.06648	.28928
1.63	1.4429	1.2750	.22501	.34458	.65301	1.06798	.28690
1.64	1.4487	1.2835	.22168	.34093	.65023	1.06948	.28454
1.65	1.4544	1.2922	.21839	.33731	.64746	1.07098	.28221
1.66	1.4601	1.3010	.21515	.33372	.64470	1.07249	.27991
1.67	1.4657	1.3099	.21195	.33016	.64194	1.07399	.27764
1.68	1.4713	1.3190	.20879	.32664	.63919	1.07550	.27540
1.69	1.4769	1.3282	.20567	.32315	.63645	1.07701	.27318
1.70	1.4825	1.3376	.20259	.31969	.63372	1.07851	.27099
1.71	1.4880	1.3471	.19955	.31626	.63099	1.08002	.26882
1.72	1.4935	1.3567	.19656	.31286	.62827	1.08152	.26668
1.73	1.4989	1.3665	.19361	.30950	.62556	1.08302	.26457
1.74	1.5043	1.3764	.19070	.30617	.62286	1.08453	.26248
1.75	1.5097	1.3865	.18782	.30287	.62016	1.08603	.26042
1.76	1.5150	1.3967	.18499	.29959	.61747	1.08753	.25838
1.77	1.5203	1.4071	.18220	.29635	.61479	1.08903	.25636
1.78	1.5256	1.4176	.17944	.29314	.61211	1.09053	.25436
1.79	1.5308	1.4282	.17672	.28997	.60945	1.09202	.25239

M	M^*	$\frac{A}{A^*}$	$\frac{p}{p_o}$	$\frac{\rho}{\rho_o}$	$\frac{T}{T_o}$	$\frac{F}{F^*}$	$\left(\frac{A}{A^*}\right)\left(\frac{p}{p_o}\right)$
1.80	1.5360	1.4390	.17404	.28682	.60680	1.09352	.25044
1.81	1.5412	1.4499	.17140	.28370	.60415	1.09500	.24851
1.82	1.5463	1.4610	.16879	.28061	.60151	1.09649	.24660
1.83	1.5514	1.4723	.16622	.27756	.59888	1.09798	.24472
1.84	1.5564	1.4837	.16369	.27453	.59626	1.09946	.24286
1.85	1.5614	1.4952	.16120	.27153	.59365	1.1009	.24102
1.86	1.5664	1.5069	.15874	.26857	.59105	1.1024	.23919
1.87	1.5714	1.5188	.15631	.26563	.58845	1.1039	.23739
1.88	1.5763	1.5308	.15392	.26272	.58586	1.1054	.23561
1.89	1.5812	1.5429	.15156	.25984	.58329	1.1068	.23385
1.90	1.5861	1.5552	.14924	.25699	.58072	1.1083	.23211
1.91	1.5909	1.5677	.14695	.25417	.57816	1.1097	.23039
1.92	1.5957	1.5804	.14469	.25138	.57561	1.1112	.22868
1.93	1.6005	1.5932	.14247	.24862	.57307	1.1126	.22699
1.94	1.6052	1.6062	.14028	.24588	.57054	1.1141	.22532
1.95	1.6099	1.6193	.13813	.24317	.56802	1.1155	.22367
1.96	1.6146	1.6326	.13600	.24049	.56551	1.1170	.22204
1.97	1.6193	1.6461	.13390	.23784	.56301	1.1184	.22042
1.98	1.6239	1.6597	.13184	.23522	.56051	1.1198	.21882
1.99	1.6285	1.6735	.12981	.23262	.55803	1.1213	.21724
2.00	1.6330	1.6875	.12780	.23005	.55556	1.1227	.21567
2.01	1.6375	1.7017	.12583	.22751	.55310	1.1241	.21412
2.02	1.6420	1.7160	.12389	.22499	.55064	1.1255	.21259
2.03	1.6465	1.7305	.12198	.22250	.54819	1.1269	.21107
2.04	1.6509	1.7452	.12009	.22004	.54576	1.1283	.20957
2.05	1.6553	1.7600	.11823	.21760	.54333	1.1297	.20808
2.06	1.6597	1.7750	.11640	.21519	.54091	1.1311	.20661
2.07	1.6640	1.7902	.11460	.21281	.53850	1.1325	.20515
2.08	1.6683	1.8056	.11282	.21045	.53611	1.1329	.20371
2.09	1.6726	1.8212	.11107	.20811	.53373	1.1352	.20228

M	M^*	$\frac{A}{A^*}$	$\frac{p}{p_o}$	$\frac{\rho}{\rho_o}$	$\frac{T}{T_o}$	$\frac{F}{F^*}$	$\left(\frac{A}{A^*}\right)\left(\frac{p}{p_o}\right)$
2.10	1.6769	1.8369	.10935	.20580	.53135	1.1366	.20087
2.11	1.6811	1.8529	.10766	.20352	.52898	1.1380	.19947
2.12	1.6853	1.8690	.10599	.20126	.52663	1.1393	.19809
2.13	1.6895	1.8853	.10434	.19902	.52428	1.1407	.19672
2.14	1.6936	1.9018	.10272	.19681	.52194	1.1420	.19537
2.15	1.6977	1.9185	.10113	.19463	.51962	1.1434	.19403
2.16	1.7018	1.9354	.09956	.19247	.51730	1.1447	.19270
2.17	1.7059	1.9525	.09802	.19033	.51499	1.1460	.19138
2.18	1.7099	1.9698	.09650	.18821	.51269	1.1474	.19008
2.19	1.7139	1.9873	.09500	.18612	.51041	1.1487	.18879
2.20	1.7179	2.0050	.09352	.18405	.50813	1.1500	.18751
2.21	1.7219	2.0229	.09207	.18200	.50586	1.1513	.18624
2.22	1.7258	2.0409	.09064	.17998	.50361	1.1526	.18499
2.23	1.7297	2.0592	.08923	.17798	.50136	1.1539	.18375
2.24	1.7336	2.0777	.08784	.17600	.49912	1.1552	.18252
2.25	1.7374	2.0964	.08648	.17404	.49689	1.1565	.18130
2.26	1.7412	2.1154	.08514	.17211	.49468	1.1578	.18009
2.27	1.7450	2.1345	.08382	.17020	.49247	1.1590	.17890
2.28	1.7488	2.1538	.08252	.16830	.49027	1.1603	.17772
2.29	1.7526	2.1734	.08123	.16643	.48809	1.1616	.17655
2.30	1.7563	2.1931	.07997	.16458	.48591	1.1629	.17539
2.31	1.7600	2.2131	.07873	.16275	.48374	1.1641	.17424
2.32	1.7637	2.2333	.07751	.16095	.48158	1.1653	.17310
2.33	1.7673	2.2537	.07631	.15916	.47944	1.1666	.17197
2.34	1.7709	2.2744	.07513	.15739	.47730	1.1678	.17085
2.35	1.7745	2.2953	.07396	.15564	.47517	1.1690	.16975
2.36	1.7781	2.3164	.07281	.15391	.47305	1.1703	.16866
2.37	1.7817	2.3377	.07168	.15220	.47095	1.1715	.16757
2.38	1.7852	2.3593	.07057	.15052	.46885	1.1727	.16649
2.39	1.7887	2.3811	.06948	.14885	.46676	1.1739	.16543

M	M^*	$\frac{A}{A^*}$	$\frac{p}{p_o}$	$\frac{\rho}{\rho_o}$	$\frac{T}{T_o}$	$\frac{F}{F^*}$	$\left(\frac{A}{A^*}\right)\left(\frac{p}{p_o}\right)$
2.40	1.7922	2.4031	.06840	.14720	.46468	1.1751	.16437
2.41	1.7957	2.4254	.06734	.14557	.46262	1.1763	.16332
2.42	1.7991	2.4479	.06630	.14395	.46056	1.1775	.16229
2.43	1.8025	2.4706	.06527	.14235	.45851	1.1786	.16126
2.44	1.8059	2.4936	.06426	.14078	.45647	1.1798	.16024
2.45	1.8093	2.5168	.06327	.13922	.45444	1.1810	.15923
2.46	1.8126	2.5403	.06229	.13768	.45242	1.1821	.15823
2.47	1.8159	2.5640	.06133	.13616	.45041	1.1833	.15724
2.48	1.8192	2.5880	.06038	.13465	.44841	1.1844	.15624
2.49	1.8225	2.6122	.05945	.13316	.44642	1.1856	.15528
2.50	1.8258	2.6367	0.5853	.13169	.44444	1.1867	.15432
2.51	1.8290	2.6615	0.5763	.13023	.44247	1.1879	.15337
2.52	1.8322	2.6865	0.5674	.12879	.44051	1.1890	.15242
2.53	1.8354	2.7117	0.5586	.12737	.43856	1.1901	.15148
2.54	1.8386	2.7372	0.5500	.12597	.43662	1.1912	.15055
2.55	1.8417	2.7630	.05415	.12458	.43469	1.1923	.14963
2.56	1.8448	2.7891	.05332	.12321	.43277	1.1934	.14871
2.57	1.8479	2.8154	.05250	.12185	.43085	1.1945	.14780
2.58	1.8510	2.8420	.05169	.12051	.42894	1.1956	.14691
2.59	1.8541	2.8689	.05090	.11418	.42705	1.1967	.14601
2.60	1.8572	2.8960	.05012	.11787	.42517	1.1978	.14513
2.61	1.8602	2.9234	.04935	.11658	.42330	1.1989	.14426
2.62	1.8632	2.9511	.04859	.11530	.42143	1.2000	.14339
2.63	1.8662	2.9791	.04784	.11403	.41957	1.2011	.14253
2.64	1.8692	3.0074	.04711	.11278	.41772	1.2021	.14168
2.65	1.8721	3.0359	.04639	.11154	.41589	1.2031	.14083
2.66	1.8750	3.0647	.04568	.11032	.41406	1.2042	.13999
2.67	1.8779	3.0938	.04498	.10911	.41224	1.2052	.13916
2.68	1.8808	3.1233	.04429	.10792	.41043	1.2062	.13834
2.69	1.8837	3.1530	.04361	.10674	.40863	1.2073	.13752

M	M^*	$\frac{A}{A^*}$	$\frac{p}{p_o}$	$\frac{\rho}{\rho_o}$	$\frac{T}{T_o}$	$\frac{F}{F^*}$	$\left(\frac{A}{A^*}\right)\left(\frac{p}{p_o}\right)$
2.70	1.8865	3.1830	.04295	.10557	.40684	1.2083	.13671
2.71	1.8894	3.2133	.04230	.10442	.40505	1.2093	.13591
2.72	1.8922	3.2440	.04166	.10328	.40327	1.2103	.13511
2.73	1.8950	3.2749	.04102	.10215	.40151	1.2113	.13432
2.74	1.8978	3.3061	.04039	.10104	.39976	1.2123	.13354
2.75	1.9005	3.3376	.03977	.09994	.39801	1.2133	.13276
2.76	1.9032	3.3695	.03917	.09885	.39627	1.2143	.13199
2.77	1.9060	3.4017	.03858	.09777	.39454	1.2153	.13123
2.78	1.9087	3.4342	.03800	.09671	.39282	1.2163	.13047
2.79	1.9114	3.4670	.03742	.09566	.39111	1.2173	.12972
2.80	1.9140	3.5001	.03685	.09462	.38941	1.2182	.12897
2.81	1.9167	3.5336	.03629	.09360	.38771	1.2192	.12823
2.82	1.9193	3.5674	.03574	.09259	.38603	1.2202	.12750
2.83	1.9220	3.6015	.03520	.09158	.38435	1.2211	.12678
2.84	1.9246	3.6359	.03467	.09059	.38268	1.2221	.12605
2.85	1.9271	3.6707	.03415	.08962	.38102	1.2230	.12534
2.86	1.9297	3.7058	.03363	.08865	.37937	1.2240	.12463
2.87	1.9322	3.7413	.03312	.08769	.37773	1.2249	.12393
2.88	1.9348	3.7771	.03262	.08674	.37610	1.2258	.12323
2.89	1.9373	3.8133	.03213	.08581	.37448	1.2268	.12254
2.90	1.9398	3.8498	.03165	.08489	.37286	1.2277	.12185
2.91	1.9423	3.8868	.03118	.08398	.37125	1.2286	.12117
2.92	1.9448	3.9238	.03071	.08308	.36965	1.2295	.12049
2.93	1.9472	3.9614	.03025	.08218	.36806	1.2304	.11982
2.94	1.9497	3.9993	.02980	.08130	.36648	1.2313	.11916
2.95	1.9521	4.0376	.02935	.08043	.36490	1.2322	.11850
2.96	1.9545	4.0763	.02891	.07957	.36333	1.2331	.11785
2.97	1.9569	4.1153	.02848	.07872	.36177	1.2340	.11720
2.98	1.9593	4.1547	.02805	.07788	.36022	1.2348	.11656
2.99	1.9616	4.1944	.02764	.07705	.35868	1.2357	.11591

M	M^*	$\frac{A}{A^*}$	$\frac{p}{p_o}$	$\frac{\rho}{\rho_o}$	$\frac{T}{T_o}$	$\frac{F}{F^*}$	$\left(\frac{A}{A^*}\right)\left(\frac{p}{p_o}\right)$
3.00	1.9640	4.2346	.02722	.07623	.35714	1.2366	.11528
3.10	1.9866	4.6573	.02345	.06852	.34223	1.2450	.10921
3.20	2.0079	5.1210	.02023	.06165	.32808	1.2530	.10359
3.30	2.0279	5.6287	.01748	.05554	.31466	1.2605	.09837
3.40	2.0466	6.1837	.01512	.05009	.30193	1.2676	.09353
3.50	2.0642	6.7896	.01311	.04523	.28986	1.2743	.08902
3.60	2.0808	7.4501	.01138	.04089	.27840	1.2807	.08482
3.70	2.0964	8.1691	.00990	.03702	.26752	1.2867	.08090
3.80	2.1111	8.9506	.00863	.03355	.25720	1.2924	.07723
3.90	2.1250	9.7990	.00753	.03044	.24740	1.2978	.07380
4.00	2.1381	10.719	.00658	.02766	.23810	1.3029	.07059
4.10	2.1505	11.715	.00577	.02516	.22925	1.3077	.06758
4.20	2.1622	12.792	.00506	.02292	.22085	1.3123	.06475
4.30	2.1732	13.955	.00445	.02090	.21286	1.3167	.06209
4.40	2.1837	15.210	.00392	.01909	.20525	1.3208	.05959
4.50	2.1936	16.562	.00346	.01745	.19802	1.3247	.05723
4.60	2.2030	18.018	.00305	.01597	.19113	1.3284	.05500
4.70	2.2119	19.583	.00270	.01463	.18457	1.3320	.05289
4.80	2.2204	21.264	.00240	.01343	.17832	1.3354	.05091
4.90	2.2284	23.067	.00213	.01233	.17235	1.3386	.04904
5.0	2.2361	25.000	$189(10)^{-5}$	.01134	.16667	1.3416	.04725
6.0	2.2953	53.180	$633(10)^{-6}$	.00519	.12195	1.3655	.03368
7.0	2.3333	104.143	$242(10)^{-6}$	.00261	.09259	1.3810	.02516
8.0	2.3591	190.109	$102(10)^{-6}$	.00141	.07246	1.3915	.01947
9.0	2.3772	327.189	$474(10)^{-7}$	.000815	.05814	1.3989	.01550
10.00	2.3904	535.938	$236(10)^{-7}$	.000495	.04762	1.4044	.01263
∞	2.4495	∞	0	0	0	1.4289	0

Table-2 RAYLEIGH LINE — One dimensional compressible flow functions for stagnation temperature change in the absence of friction and area change.

For a perfect gas with constant specific heat and molecular weight k = 1.4.

M	$\frac{T_o}{T_o^*}$	$\frac{T}{T^*}$	$\frac{p}{p^*}$	$\frac{p_o}{p_o^*}$	$\frac{V}{V^*}$
0	0	0	2.4000	1.2679	0
0.01	.000480	.000576	2.3997	1.2678	.000240
.02	.00192	.00230	2.3987	1.2675	.000959
.03	.00431	.00516	2.3970	1.2671	.00216
.04	.00765	.00917	2.3946	1.2665	.00383
.05	.01192	.01430	2.3916	1.2657	.00598
.06	.01712	.02053	2.3880	1.2647	.00860
.07	.02322	.02784	2.3837	1.2636	.01168
.08	.03021	.03621	2.3787	1.2623	.01522
.09	.03807	.04562	2.3731	1.2608	.01922
.10	.04678	.05602	2.3669	1.2591	.02367
.11	.05630	.06739	2.3600	1.2573	.02856
.12	.06661	.07970	2.3526	1.2554	.03388
.13	.07768	.09290	2.3445	1.2533	.03962
.14	.08947	.10695	2.3359	1.2510	.04578
.15	.10196	.12181	2.3267	1.2486	.05235
.16	.11511	.13743	2.3170	1.2461	.05931
.17	.12888	.15377	2.3067	1.2434	.06666
.18	.14324	.17078	2.2959	1.2406	.07438
.19	.15814	.18841	2.2845	1.2377	.08247
.20	.17355	.20661	2.2727	1.2346	.09091
.21	.18943	.22533	2.2604	1.2314	.09969
.22	.20574	.24452	2.2477	1.2281	.10879
.23	.22244	.26413	2.2345	1.2248	.11820
.24	.23948	.28411	2.2209	1.2213	.12792

M	$\frac{T_o}{T_o^*}$	$\frac{T}{T^*}$	$\frac{p}{p^*}$	$\frac{p_o}{p_o^*}$	$\frac{V}{V^*}$
.25	.25684	.30440	2.2069	1.2177	.13793
.26	.27446	.32496	2.1925	1.2140	.14821
.27	.29231	.34573	2.1777	1.2102	.15876
.28	.31035	.36667	2.1626	1.2064	.16955
.29	.32855	.38773	2.1472	1.2025	.18058
.30	.34686	.40887	2.1314	1.1985	.19183
.31	.36525	.43004	2.1154	1.1945	.20329
.32	.38369	.45119	2.0991	1.1904	.21494
.33	.40214	.47228	2.0825	1.1863	.22678
.34	.42057	.49327	2.0657	1.1821	.23879
.35	.43894	.51413	2.0487	1.1779	.25096
.36	.45723	.53482	2.0314	1.1737	.26327
.37	.47541	.55530	2.0140	1.1695	.27572
.38	.49346	.57553	1.9964	1.1652	.28828
.39	.51134	.59549	1.9787	1.1609	.30095
0.40	.52903	.61515	1.9608	1.1566	.31372
.41	.54651	.63448	1.9428	1.1523	.32658
.42	.56376	.65345	1.9247	1.1480	.33951
.43	.58075	.67205	1.9065	1.1437	.35251
.44	.59748	.69025	1.8882	1.1394	.36556
.45	.61393	.70803	1.8699	1.1351	.37865
.46	.63007	.72538	1.8515	1.1308	.39178
.47	.64589	.74228	1.8331	1.1266	.40493
.48	.66139	.75871	1.8147	1.1224	.41810
.49	.67655	.77466	1.7962	1.1182	.43127
.50	.69136	.79012	1.7778	1.1140	.44445
.51	.70581	.80509	1.7594	1.1099	.45761
.52	.71990	.81955	1.7410	1.1059	.47075
.53	.73361	.83351	1.7226	1.1019	.48387
.54	.74695	.84695	1.7043	1.0979	.49696

M	$\frac{T_o}{T_o^*}$	$\frac{T}{T^*}$	$\frac{p}{p^*}$	$\frac{p_o}{p_o^*}$	$\frac{V}{V^*}$
.55	.75991	.85987	1.6860	1.09397	.51001
.56	.77248	.87227	1.6678	1.09010	.52302
.57	.78467	.88415	1.6496	1.08630	.53597
.58	.79647	.89552	1.6316	1.08255	.54887
.59	.80789	.90637	1.6136	1.07887	.56170
.60	.81892	.91670	1.5957	1.07525	.57447
.61	.82956	.92653	1.5780	1.07170	.58716
.62	.83982	.93585	1.5603	1.06821	.59978
.63	.84970	.94466	1.5427	1.06480	.61232
.64	.85920	.95298	1.5253	1.06146	.62477
.65	.86833	.96081	1.5080	1.05820	.63713
.66	.87709	.96816	1.4908	1.05502	.64941
.67	.88548	.97503	1.4738	1.05192	.66159
.68	.89350	.98144	1.4569	1.04890	.67367
.69	.90117	.98739	1.4401	1.04596	.68564
.70	.90850	.99289	1.4235	1.04310	.69751
.71	.91548	.99796	1.4070	1.04033	.70927
.72	.92212	1.00260	1.3907	1.03764	.72093
.73	.92843	1.00682	1.3745	1.03504	.73248
.74	.93442	1.01062	1.3585	1.03253	.74392
.75	.94009	1.01403	1.3427	1.03010	.75525
.76	.94546	1.01706	1.3270	1.02776	.76646
.77	.95052	1.01971	1.3115	1.02552	.77755
.78	.95528	1.02198	1.2961	1.02337	.78852
.79	.95975	1.02390	1.2809	1.02131	.79938
0.80	.96394	1.02548	1.2658	1.01934	.81012
.81	.96786	1.02672	1.2509	1.01746	.82075
.82	.97152	1.02763	1.2362	1.01569	.83126
.83	.97492	1.02823	1.2217	1.01399	.84164
.84	.97807	1.02853	1.2073	1.01240	.85190

M	$\frac{T_o}{T_o^*}$	$\frac{T}{T^*}$	$\frac{p}{p^*}$	$\frac{p_o}{p_o^*}$	$\frac{V}{V^*}$
.85	.98097	1.02854	1.1931	1.01091	.86204
.86	.98363	1.02826	1.1791	1.00951	.87206
.87	.98607	1.02771	1.1652	1.00819	.88196
.88	.98828	1.02690	1.1515	1.00698	.89175
.89	.99028	1.02583	1.1380	1.00587	.90142
.90	.99207	1.02451	1.1246	1.00485	.91097
.91	.99366	1.02297	1.1114	1.00393	.92039
.92	.99506	1.02120	1.09842	1.00310	.92970
.93	.99627	1.01921	1.08555	1.00237	.93889
.94	.99729	1.01702	1.07285	1.00174	.94796
.95	.99814	1.01463	1.06030	1.00121	.95692
.96	.99883	1.01205	1.04792	1.00077	.96576
.97	.99935	1.00929	1.03570	1.00043	.97449
.98	.99972	1.00636	1.02364	1.00019	.98311
.99	.99993	1.00326	1.01174	1.00004	.99161
1.00	1.00000	1.00000	1.00000	1.00000	1.00000
1.01	.99993	.99659	.98841	1.00004	1.00828
1.02	.99973	.99304	.97697	1.00019	1.01644
1.03	.99940	.98936	.96569	1.00043	1.02450
1.04	.99895	.98553	.95456	1.00077	1.03246
1.05	.99838	.98161	.94358	1.00121	1.04030
1.06	.99769	.97755	.93275	1.00175	1.04804
1.07	.99690	.97339	.92206	1.00238	1.05567
1.08	.99600	.96913	.91152	1.00311	1.06320
1.09	.99501	.96477	.90112	1.00394	1.07062
1.10	.99392	.96031	.89086	1.00486	1.07795
1.11	.99274	.95577	.88075	1.00588	1.08518
1.12	.99148	.95115	.87078	1.00699	1.09230
1.13	.99013	.94646	.86094	1.00820	1.09933
1.14	.98871	.94169	.85123	1.00951	1.10626

M	$\frac{T_o}{T_o^*}$	$\frac{T}{T^*}$	$\frac{p}{p^*}$	$\frac{p_o}{p_o^*}$	$\frac{V}{V^*}$
1.15	.98721	.93685	.84166	1.01092	1.1131
1.16	.98564	.93195	.83222	1.01243	1.1198
1.17	.98400	.92700	.82292	1.01403	1.1264
1.18	.98230	.92200	.81374	1.01572	1.1330
1.19	.98054	.91695	.80468	1.01752	1.1395
1.20	.97872	.91185	.79576	1.01941	1.1459
1.21	.97685	.90671	.78695	1.02140	1.1522
1.22	.97492	.90153	.77827	1.02348	1.1584
1.23	.97294	.89632	.76971	1.02566	1.1645
1.24	.97092	.89108	.76127	1.02794	1.1705
1.25	.96886	.88581	.75294	1.03032	1.1764
1.26	.96675	.88052	.74473	1.03280	1.1823
1.27	.96461	.87521	.73663	1.03536	1.1881
1.28	.96243	.86988	.72865	1.03803	1.1938
1.29	.96022	.86453	.72078	1.04080	1.1994
1.30	.95798	.85917	.71301	1.04365	1.2050
1.31	.95571	.85380	.70535	1.04661	1.2105
1.32	.95341	.84843	.69780	1.04967	1.2159
1.33	.95108	.84305	.69035	1.05283	1.2212
1.34	.94873	.83766	.68301	1.05608	1.2264
1.35	.94636	.83227	.67577	1.05943	1.2316
1.36	.94397	.82698	.66863	1.06288	1.2367
1.37	.94157	.82151	.66159	1.06642	1.2417
1.38	.93915	.81613	.65464	1.07006	1.2467
1.39	.93671	.81076	.64778	1.07380	1.2516
1.40	.93425	.80540	.64102	1.07765	1.2564
1.41	.93178	.80004	.63436	1.08159	1.2612
1.42	.92931	.79469	.62779	1.08563	1.2659
1.43	.92683	.78936	.62131	1.08977	1.2705
1.44	.92434	.78405	.61491	1.09400	1.2751

M	$\frac{T_o}{T_o^*}$	$\frac{T}{T^*}$	$\frac{p}{p^*}$	$\frac{p_o}{p_o^*}$	$\frac{V}{V^*}$
1.45	.92184	.77875	.60860	1.0983	1.2796
1.46	.91933	.77346	.60237	1.1028	1.2840
1.47	.91682	.76819	.59623	1.1073	1.2884
1.48	.91431	.76294	.59018	1.1120	1.2927
1.49	.91179	.75771	.58421	1.1167	1.2970
1.50	.90928	.75250	.57831	1.1215	1.3012
1.51	.90676	.74731	.57250	1.1264	1.3054
1.52	.90424	.74215	.56677	1.1315	1.3095
1.53	.90172	.73701	.56111	1.1367	1.3135
1.54	.89920	.73189	.55553	1.1420	1.3175
1.55	.89669	.72680	.55002	1.1473	1.3214
1.56	.89418	.72173	.54458	1.1527	1.3253
1.57	.89167	.71669	.53922	1.1582	1.3291
1.58	.88917	.71168	.53393	1.1639	1.3329
1.59	.88668	.70669	.52871	1.1697	1.3366
1.60	.88419	.70173	.52356	1.1756	1.3403
1.61	.88170	.69680	.51848	1.1816	1.3439
1.62	.87922	.69190	.51346	1.1877	1.3475
1.63	.87675	.68703	.50851	1.1939	1.3511
1.64	.87429	.68219	.50363	1.2002	1.3546
1.65	.87184	.67738	.49881	1.2066	1.3580
1.66	.86940	.67259	.49405	1.2131	1.3614
1.67	.86696	.66784	.48935	1.2197	1.3648
1.68	.86453	.66312	.48471	1.2264	1.3681
1.69	.86211	.65843	.48014	1.2332	1.3713
1.70	.85970	.65377	.47563	1.2402	1.3745
1.71	.85731	.64914	.47117	1.2473	1.3777
1.72	.85493	.64455	.46677	1.2545	1.3809
1.73	.85256	.63999	.46242	1.2618	1.3840
1.74	.85020	.63546	.45813	1.2692	1.3871

M	$\frac{T_o}{T_o^*}$	$\frac{T}{T^*}$	$\frac{p}{p^*}$	$\frac{p_o}{p_o^*}$	$\frac{V}{V^*}$
1.75	.84785	.63096	.45390	1.2767	1.3901
1.76	.84551	.62649	.44972	1.2843	1.3931
1.77	.84318	.62205	.44559	1.2920	1.3960
1.78	.84087	.61765	.44152	1.2998	1.3989
1.79	.83857	.61328	.43750	1.3078	1.4018
1.80	.83628	.60894	.43353	1.3159	1.4046
1.81	.83400	.60463	.42960	1.3241	1.4074
1.82	.83174	.60036	.42573	1.3324	1.4102
1.83	.82949	.59612	.42191	1.3408	1.4129
1.84	.82726	.59191	.41813	1.3494	1.4156
1.85	.82504	.58773	.41440	1.3581	1.4183
1.86	.82283	.58359	.41072	1.3669	1.4209
1.87	.82064	.57948	.40708	1.3758	1.4235
1.88	.81846	.57540	.40349	1.3848	1.4261
1.89	.81629	.57135	.39994	1.3940	1.4286
1.90	.81414	.56734	.39643	1.4033	1.4311
1.91	.81200	.56336	.39297	1.4127	1.4336
1.92	.80987	.55941	.38955	1.4222	1.4360
1.93	.80776	.55549	.38617	1.4319	1.4384
1.94	.80567	.55160	.38283	1.4417	1.4408
1.95	.80359	.54774	.37954	1.4516	1.4432
1.96	.80152	.54391	.37628	1.4616	1.4455
1.97	.79946	.54012	.37306	1.4718	1.4478
1.98	.79742	.53636	.36988	1.4821	1.4501
1.99	.79540	.53263	.36674	1.4925	1.4523
2.00	.79339	.52893	.36364	1.5031	1.4545
2.01	.79139	.52526	.36057	1.5138	1.4567
2.02	.78941	.52161	.35754	1.5246	1.4589
2.03	.78744	.51800	.35454	1.5356	1.4610
2.04	.78549	.51442	.35158	1.5467	1.4631

M	$\frac{T_o}{T_o^*}$	$\frac{T}{T^*}$	$\frac{p}{p^*}$	$\frac{p_o}{p_o^*}$	$\frac{V}{V^*}$
2.05	.78355	.51087	.34866	1.5579	1.4652
2.06	.78162	.50735	.34577	1.5693	1.4673
2.07	.77971	.50386	.34291	1.5808	1.4694
2.08	.77781	.50040	.34009	1.5924	1.4714
2.09	.77593	.49697	.33730	1.6042	1.4734
2.10	.77406	.49356	.33454	1.6161	1.4753
2.11	.77221	.49018	.33181	1.6282	1.4773
2.12	.77037	.48683	.32912	1.6404	1.4792
2.13	.76854	.48351	.32646	1.6528	1.4811
2.14	.76673	.48022	.32383	1.6653	1.4830
2.15	.76493	.47696	.32122	1.6780	1.4849
2.16	.76314	.47373	.31864	1.6908	1.4867
2.17	.76137	.47052	.31610	1.7037	1.4885
2.18	.75961	.46734	.31359	1.7168	1.4903
2.19	.75787	.46419	.31110	1.7300	1.4921
2.20	.75614	.46106	.30864	1.7434	1.4939
2.21	.75442	.45796	.30621	1.7570	1.4956
2.22	.75271	.45489	.30381	1.7707	1.4973
2.23	.75102	.45184	.30143	1.7846	1.4990
2.24	.74934	.44882	.29908	1.7986	1.5007
2.25	.74767	.44582	.29675	1.8128	1.5024
2.26	.74602	.44285	.29445	1.8271	1.5040
2.27	.74438	.43990	.29218	1.8416	1.5056
2.28	.74275	.43698	.28993	1.8562	1.5072
2.29	.74114	.43409	.28771	1.8710	1.5088
2.30	.73954	.43122	.28551	1.8860	1.5104
2.31	.73795	.42837	.28333	1.9012	1.5119
2.32	.73638	.42555	.28118	1.9165	1.5134
2.33	.73482	.42276	.27905	1.9320	1.5150
2.34	.73327	.41999	.27695	1.9476	1.5165

M	$\frac{T_o}{T_o^*}$	$\frac{T}{T^*}$	$\frac{p}{p^*}$	$\frac{p_o}{p_o^*}$	$\frac{V}{V^*}$
2.35	.73173	.41724	.27487	1.9634	1.5180
2.36	.73020	.41451	.27281	1.9794	1.5195
2.37	.72868	.41181	.27077	1.9955	1.5209
2.38	.72718	.40913	.26875	2.0118	1.5223
2.39	.72569	.40647	.26675	2.0283	1.5237
2.40	.72421	.40383	.26478	2.0450	1.5252
2.41	.72274	.40122	.26283	2.0619	1.5266
2.42	.72129	.39863	.26090	2.0789	1.5279
2.43	.71985	.39606	.25899	2.0961	1.5293
2.44	.71842	.39352	.25710	2.1135	1.5306
2.45	.71700	.39100	.25523	2.1311	1.5320
2.46	.71559	.38850	.25337	2.1489	1.5333
2.47	.71419	.38602	.25153	2.1669	1.5346
2.48	.71280	.38356	.24972	2.1850	1.5359
2.49	.71142	.38112	.24793	2.2033	1.5372
2.50	.71005	.37870	.24616	2.2218	1.5385
2.51	.70870	.37630	.24440	2.2405	1.5398
2.52	.70736	.37392	.24266	2.2594	1.5410
2.53	.70603	.37157	.24094	2.2785	1.5422
2.54	.70471	.36923	.23923	2.2978	1.5434
2.55	.70340	.36691	.23754	2.3173	1.5446
2.56	.70210	.36461	.23587	2.3370	1.5458
2.57	.70081	.36233	.23422	2.3569	1.5470
2.58	.69953	.36007	.23258	2.3770	1.5482
2.59	.69825	.35783	.23096	2.3972	1.5494
2.60	.69699	.35561	.22936	2.4177	1.5505
2.61	.69574	.35341	.22777	2.4384	1.5516
2.62	.69450	.35123	.22620	2.4593	1.5527
2.63	.69327	.34906	.22464	2.4804	1.5538
2.64	.69205	.34691	.22310	2.5017	1.5549

M	$\frac{T_o}{T_o^*}$	$\frac{T}{T^*}$	$\frac{p}{p^*}$	$\frac{p_o}{p_o^*}$	$\frac{V}{V^*}$
2.65	.69084	.34478	.22158	2.5233	1.5560
2.66	.68964	.34267	.22007	2.5451	1.5571
2.67	.68845	.34057	.21857	2.5671	1.5582
2.68	.68727	.33849	.21709	2.5802	1.5593
2.69	.68610	.33643	.21562	2.6116	1.5603
2.70	.68494	.33439	.21417	2.6342	1.5613
2.71	.68378	.33236	.21273	2.6571	1.5623
2.72	.68283	.33035	.21131	2.6802	1.5633
2.73	.68150	.32836	.20990	2.7035	1.5644
2.74	.68038	.32638	.20850	2.7270	1.5654
2.75	.67926	.32442	.20712	.27508	1.5663
2.76	.67815	.32248	.20575	.27748	1.5673
2.77	.67704	.32055	.20439	.27990	1.5683
2.78	.67595	.31864	.20305	2.8235	1.5692
2.79	.67487	.31674	.20172	2.8482	1.5702
2.80	.67380	.31486	.20040	2.8731	1.5711
2.81	.67273	.31299	.19909	2.8982	1.5721
2.82	.67167	.31114	.19780	2.9236	1.5730
2.83	.67062	.30931	.19652	2.9493	1.5739
2.84	.66958	.30749	.19525	2.9752	1.5748
2.85	.66855	.30568	.19399	3.0013	1.5757
2.86	.66752	.30389	.19274	3.0277	1.5766
2.87	.66650	.30211	.19151	3.0544	1.5775
2.88	.66549	.30035	.19029	3.0813	1.5784
2.89	.66449	.29860	.18908	3.1084	1.5792
2.90	.66350	.29687	1.8788	3.1358	1.5801
2.91	.66252	.29515	1.8669	3.1635	1.5809
2.92	.66154	.29344	1.8551	3.1914	1.5818
2.93	.66057	.29175	1.8435	3.2196	1.5826
2.94	.65961	.29007	1.8320	3.2481	1.5834

M	$\frac{T_o}{T_o^*}$	$\frac{T}{T^*}$	$\frac{p}{p^*}$	$\frac{p_o}{p_o^*}$	$\frac{V}{V^*}$
2.95	.65865	.28841	.18205	3.2768	1.5843
2.96	.65770	.28676	.18091	3.3058	1.5851
2.97	.65676	.28512	.17978	3.3351	1.5859
2.98	.65583	.28349	.17867	3.3646	1.5867
2.99	.65490	.28188	.17757	3.3944	1.5875
3.00	.65398	.28028	.17647	3.4244	1.5882
3.50	.61580	.21419	.13223	5.3280	1.6198
4.00	.58909	.16831	.10256	8.2268	1.6410
4.50	.56983	.13540	.08177	12.502	1.6559
5.00	.55555	.11111	.06667	18.634	1.6667
6.00	.53633	.07849	.04669	38.946	1.6809
7.00	.52437	.05826	.03448	75.414	1.6896
8.00	.51646	.04491	.02649	136.62	1.6954
9.00	.51098	.03565	.02098	233.88	1.6993
10.00	.50702	.02897	.01702	381.62	1.7021
∞	.48980	0	0	∞	1.7143

Table-3 FANNO LINE — One Dimensional compressible flow functions for Adiabatic flow at constant Area with Friction.

For a perfect gas with constant specific heat and molecular weight k = 1.4.

M	$\frac{T}{T^*}$	$\frac{p}{p^*}$	$\frac{p_o}{p_o^*}$	$\frac{V}{V^*}$	$\frac{F}{F^*}$	$\frac{\lambda_{av} L_{max}}{D}$
0	1.2000	∞	∞	0	∞	∞
0.01	1.2000	109.544	57.874	.01095	45.650	7134.40
.02	1.1999	54.770	28.942	.02191	22.834	1778.45
.03	1.1998	36.511	19.300	.03286	15.232	787.08
.04	1.1996	27.382	14.482	.04381	11.435	440.35
.05	1.1994	21.903	11.5914	.05476	9.1584	280.02
.06	1.1991	18.251	9.6659	.06570	7.6428	193.03
.07	1.1988	15.642	8.2915	.07664	6.5620	140.66
.08	1.1985	13.684	7.2616	.08758	5.7529	106.72
.09	1.1981	12.162	6.4614	.09851	5.1249	83.496
.10	1.1976	10.9435	5.8218	.10943	4.6236	66.922
.11	1.1971	9.9465	5.2992	.12035	4.2146	54.688
.12	1.1966	9.1156	4.8643	.13126	3.8747	45.408
.13	1.1960	8.4123	4.4968	.14216	3.5880	38.207
.14	1.1953	7.8093	4.1824	.15306	3.3432	32.511
.15	1.1946	7.2866	3.9103	.16395	3.1317	27.932
.16	1.1939	6.8291	3.6727	.17482	2.9474	24.198
.17	1.1931	6.4252	3.4635	.18568	2.7855	21.115
.18	1.1923	6.0662	3.2779	.19654	2.6422	18.543
.19	1.1914	5.7448	3.1123	.20739	2.5146	16.375
.20	1.1905	5.4555	2.9635	.21822	2.4004	14.533
.21	1.1895	5.1936	2.8293	.22904	2.2976	12.956
.22	1.1885	4.9554	2.7076	.23984	2.2046	11.596
.23	1.1874	4.7378	2.5968	.25063	2.1203	10.416
.24	1.1863	4.5383	2.4956	.26141	2.0434	9.3865

M	$\frac{T}{T^*}$	$\frac{p}{p^*}$	$\frac{p_o}{p_o^*}$	$\frac{V}{V^*}$	$\frac{F}{F^*}$	$\frac{\lambda_{av} L_{max}}{D}$
.25	1.1852	4.3546	2.4027	.27217	1.9732	8.4834
.26	1.1840	4.1850	2.3173	.28291	1.9088	7.6876
.27	1.1828	4.0280	2.2385	.29364	1.8496	6.9832
.28	1.1815	3.8820	2.1656	.30435	1.7950	6.3572
.29	1.1802	3.7460	2.0979	.31504	1.7446	5.7989
.30	1.1788	3.6190	2.0351	.32572	1.6979	5.2992
.31	1.1774	3.5002	1.9765	.33637	1.6546	4.8507
.32	1.1759	3.3888	1.9219	.34700	1.6144	4.4468
.33	1.1744	3.2840	1.8708	.35762	1.5769	4.0821
.34	1.1729	3.1853	1.8229	.36822	1.5420	3.7520
.35	1.1713	3.0922	1.7780	.37880	1.5094	3.4525
.36	1.1697	3.0042	1.7358	.38935	1.4789	3.1801
.37	1.1680	2.9209	1.6961	.39988	1.4503	2.9320
.38	1.1663	2.8420	1.6587	.41039	1.4236	2.7055
.39	1.1646	2.7671	1.6234	.42087	1.3985	2.4983
0.40	1.1628	2.6958	1.5901	.43133	1.3749	2.3085
.41	1.1610	2.6280	1.5587	.44177	1.3527	2.1344
.42	1.1591	2.5634	1.5289	.45218	1.3318	1.9744
.43	1.1572	2.5017	1.5007	.46257	1.3122	1.8272
.44	1.1553	2.4428	1.4739	.47293	1.2937	1.6915
.45	1.1533	2.3865	1.4486	.48326	1.2763	1.5664
.46	1.1513	2.3326	1.4246	.49357	1.2598	1.4509
.47	1.1492	2.2809	1.4018	.50385	1.2443	1.3442
.48	1.1471	2.2314	1.3801	.51410	1.2296	1.2453
.49	1.1450	2.1838	1.3595	.52433	1.2158	1.1539
.50	1.1429	2.1381	1.3399	.53453	1.2027	1.06908
.51	1.1407	2.0942	1.3212	.54469	1.1903	.99042
.52	1.1384	2.0519	1.3034	.55482	1.1786	.91741
.53	1.1362	2.0112	1.2864	.56493	1.1675	.84963
.54	1.1339	1.9719	1.2702	.57501	1.1571	.78662

M	$\frac{T}{T^*}$	$\frac{p}{p^*}$	$\frac{p_o}{p_o^*}$	$\frac{V}{V^*}$	$\frac{F}{F^*}$	$\frac{\lambda_{av} L_{\max}}{D}$
.55	1.1315	1.9341	1.2549	.58506	1.1472	.72805
.56	1.1292	1.8976	1.2403	.59507	1.1378	.67357
.57	1.1268	1.8623	1.2263	.60505	1.1289	.62286
.58	1.1244	1.8282	1.2130	.61500	1.1205	.57568
.59	1.1219	1.7952	1.2003	.62492	1.1126	.53174
.60	1.1194	1.7634	1.1882	.63481	1.10504	.49081
.61	1.1169	1.7325	1.1766	.64467	1.09793	.45270
.62	1.1144	1.7026	1.1656	.65449	1.09120	.41720
.63	1.1118	1.6737	1.1551	.66427	1.08485	.38411
.64	1.1091	1.6456	1.1451	.67402	1.07883	.35330
.65	1.10650	1.6183	1.1356	.68374	1.07314	.32460
.66	1.10383	1.5919	1.1265	.69342	1.06777	.29785
.67	1.10114	1.5662	1.1179	.70306	1.06271	.27295
.68	1.09842	1.5413	1.1097	.71267	1.05792	.24978
.69	1.09567	1.5170	1.1018	.72225	1.05340	.22821
.70	1.09290	1.4934	1.09436	.73179	1.04915	.20814
.71	1.09010	1.4705	1.08729	.74129	1.04514	.18949
.72	1.08727	1.4482	1.08057	.75076	1.04137	.17215
.73	1.08442	1.4265	1.07419	.76019	1.03783	.15606
.74	1.08155	1.4054	1.06815	.76958	1.03450	.14113
.75	1.07865	1.3848	1.06242	.77893	1.03137	.12728
.76	1.07573	1.3647	1.05700	.78825	1.02844	.11446
.77	1.07279	1.3451	1.05188	.79753	1.02570	.10262
.78	1.06982	1.3260	1.04705	.80677	1.02314	.09167
.79	1.06684	1.3074	1.04250	.81598	1.02075	1.08159
0.80	1.06383	1.2892	1.03823	.82514	1.01853	.07229
.81	1.06080	1.2715	1.03422	.83426	1.01646	.06375
.82	1.05775	1.2542	1.03047	.84334	1.01455	.05593
.83	1.05468	1.2373	1.02696	.85239	1.01278	.04878
.84	1.05160	1.2208	1.02370	.86140	1.01115	.04226

M	$\frac{T}{T^*}$	$\frac{p}{p^*}$	$\frac{p_o}{p_o^*}$	$\frac{V}{V^*}$	$\frac{F}{F^*}$	$\frac{\lambda_{av} L_{max}}{D}$
.85	1.04849	1.2047	1.02067	.87037	1.00966	.03632
.86	1.04537	1.1889	1.01787	.87929	1.00829	.03097
.87	1.04223	1.1735	1.01529	.88818	1.00704	.02613
.88	1.03907	1.1584	1.01294	.89703	1.00591	.02180
.89	1.03589	1.1436	1.01080	.90583	1.00490	.01793
.90	1.03270	1.12913	1.00887	.91459	1.00399	.014513
.91	1.02950	1.11500	1.00714	.92332	1.00318	.011519
.92	1.02627	1.10114	1.00560	.93201	1.00248	.008916
.93	1.02304	1.08758	1.00426	.94065	1.00188	.006694
.94	1.01978	1.07430	1.00311	.94925	1.00136	.004815
.95	1.01652	1.06129	1.00215	.95782	1.00093	.003280
.96	1.01324	1.04854	1.00137	.96634	1.00059	.002056
.97	1.00995	1.03605	1.00076	.97481	1.00033	.001135
.98	1.00664	1.02379	1.00033	.98324	1.00014	.000493
.99	1.00333	1.01178	1.00008	.99164	1.00003	.000120
1.00	1.00000	1.00000	1.00000	1.00000	1.00000	0
1.01	.99666	.98844	1.00008	1.00831	1.00003	.000114
1.02	.99331	.97711	1.00033	1.01658	1.00013	.000458
1.03	.98995	.96598	1.00073	1.02481	1.00030	.001013
1.04	.98658	.95506	1.00130	1.03300	1.00053	.001771
1.05	.98320	.94435	1.00203	1.04115	1.00082	.002712
1.06	.97982	.93383	1.00291	1.04925	1.00116	.003837
1.07	.97642	.92350	1.00394	1.05731	1.00155	.005129
1.08	.97302	.91335	1.00512	1.06533	1.00200	.006582
1.09	.96960	.90338	1.00645	1.07331	1.00250	.008185
1.10	.96618	.89359	1.00793	1.08124	1.00305	.009933
1.11	.96276	.88397	1.00955	1.08913	1.00365	.011813
1.12	.95933	.87451	1.01131	1.09698	1.00429	.013824
1.13	.95589	.86522	1.01322	1.10479	1.00497	.015949
1.14	.95244	.85608	1.01527	1.11256	1.00569	.018187

M	$\frac{T}{T^*}$	$\frac{p}{p^*}$	$\frac{p_o}{p_o^*}$	$\frac{V}{V^*}$	$\frac{F}{F^*}$	$\frac{\lambda_{av} L_{max}}{D}$
1.15	.94899	.84710	1.01746	1.1203	1.00646	.02053
1.16	.94554	.83827	1.01978	1.1280	1.00726	.02298
1.17	.94208	.82958	1.02224	1.1356	1.00810	.02552
1.18	.93862	.82104	1.02484	1.1432	1.00897	.02814
1.19	.93515	.81263	1.02757	1.1508	1.00988	.03085
1.20	.93168	.80436	1.03044	1.1583	1.01082	.03364
1.21	.92820	.79623	1.03344	1.1658	1.01178	.03650
1.22	.92473	.78822	1.03657	1.1732	1.01278	.03942
1.23	.92125	.78034	1.03983	1.1806	1.01381	.04241
1.24	.91777	.77258	1.04323	1.1879	1.01486	.04547
1.25	.91429	.76495	1.04676	1.1952	1.01594	.04858
1.26	.91080	.75743	1.05041	1.2025	1.01705	.05174
1.27	.90732	.75003	1.05419	1.2097	1.01818	.05494
1.28	.90383	.74274	1.05809	1.2169	1.01933	.05820
1.29	.90035	.73556	1.06213	1.2240	1.02050	.06150
1.30	.89686	.72848	1.06630	1.2311	1.02169	.06483
1.31	.89338	.72152	1.07060	1.2382	1.02291	.06820
1.32	.88989	.71465	1.07502	1.2452	1.02415	.07161
1.33	.88641	.70789	1.07957	1.2522	1.02540	.07504
1.34	.88292	.70123	1.08424	1.2591	1.02666	.07850
1.35	.87944	.69466	1.08904	1.2660	1.02794	.08199
1.36	.87596	.68818	1.09397	1.2729	1.02924	.08550
1.37	.87249	.68180	1.09902	1.2797	1.03056	.08904
1.38	.86901	.67551	1.10419	1.2864	1.03189	.09259
1.39	.86554	.66931	1.10948	1.2932	1.03323	.09616
1.40	.86207	.66320	1.1149	1.2999	1.03458	.09974
1.41	.85860	.65717	1.1205	1.3065	1.03595	.10333
1.42	.85514	.65122	1.1262	1.3131	1.03733	.10694
1.43	.85168	.64536	1.1320	1.3197	1.03872	.11056
1.44	.84822	.63958	1.1379	1.3262	1.04012	.11419

M	$\frac{T}{T^*}$	$\frac{p}{p^*}$	$\frac{p_o}{p_o^*}$	$\frac{V}{V^*}$	$\frac{F}{F^*}$	$\frac{\lambda_{av} L_{max}}{D}$
1.45	.84477	.63387	1.1440	1.3327	1.04153	.11782
1.46	.84133	.62824	1.1502	1.3392	1.04295	.12146
1.47	.83788	.62269	1.1565	1.3456	1.04438	.12510
1.48	.83445	.61722	1.1629	1.3520	1.04581	.12875
1.49	.83101	.61181	1.1695	1.3583	1.04725	.13240
1.50	.82759	.60648	1.1762	1.3646	1.04870	.13605
1.51	.82416	.60122	1.1830	1.3708	1.05016	.13970
1.52	.82075	.59602	1.1899	1.3770	1.05162	.14335
1.53	.81734	.59089	1.1970	1.3832	1.05309	.14699
1.54	.81394	.58583	1.2043	1.3894	1.05456	.15063
1.55	.81054	.58084	1.2116	1.3955	1.05604	.15427
1.56	.80715	.57591	1.2190	1.4015	1.05752	.15790
1.57	.80376	.57104	1.2266	1.4075	1.05900	.16152
1.58	.80038	.56623	1.2343	1.4135	1.06049	.16514
1.59	.79701	.56148	1.2422	1.4195	1.06198	.16876
1.60	.79365	.55679	1.2502	1.4254	1.06348	.17236
1.61	.79030	.55216	1.2583	1.4313	1.06498	.17595
1.62	.78695	.54759	1.2666	1.4371	1.06648	.17953
1.63	.78361	.54308	1.2750	1.4429	1.06798	.18311
1.64	.78028	.53862	1.2835	1.4487	1.06948	.18667
1.65	.77695	.53421	1.2922	1.4544	1.07098	.19022
1.66	.77363	.52986	1.3010	1.4601	1.07249	.19376
1.67	.77033	.52556	1.3099	1.4657	1.07399	.19729
1.68	.76703	.52131	1.3190	1.4713	1.07550	.20081
1.69	.76374	.51711	1.3282	1.4769	1.07701	.20431
1.70	.76046	.51297	1.3376	1.4825	1.07851	.20780
1.71	.75718	.50887	1.3471	1.4880	1.08002	.21128
1.72	.75392	.50482	1.3567	1.4935	1.08152	.21474
1.73	.75067	.50082	1.3665	1.4989	1.08302	.21819
1.74	.74742	.49686	1.3764	1.5043	1.08453	.22162

Table 3 contd..

M	$\frac{T}{T^*}$	$\frac{p}{p^*}$	$\frac{p_o}{p_o^*}$	$\frac{V}{V^*}$	$\frac{F}{F^*}$	$\frac{\lambda_{av} L_{max}}{D}$
1.75	.74419	.49295	1.3865	1.5097	1.08603	.22504
1.76	.74096	.48909	1.3967	1.5150	1.08753	.22844
1.77	.73774	.48527	1.4070	1.5203	1.08903	.23183
1.78	.73453	.48149	1.4175	1.5256	1.09053	.23520
1.79	.73134	.47776	1.4282	1.5308	1.09202	.23855
1.80	.72816	.47407	1.4390	1.5360	1.09352	.24189
1.81	.72498	.47042	1.4499	1.5412	1.09500	.24521
1.82	.72181	.46681	1.4610	1.5463	1.09649	.24851
1.83	.71865	.46324	1.4723	1.5514	1.09798	.25180
1.84	.71551	.45972	1.4837	1.5564	1.09946	.25507
1.85	.71238	.45623	1.4952	1.5614	1.1009	.25832
1.86	.70925	.45278	1.5069	1.5664	1.1024	.26156
1.87	.70614	.44937	1.5188	1.5714	1.1039	.26478
1.88	.70304	.44600	1.5308	1.5763	1.1054	.26798
1.89	.69995	.44266	1.5429	1.5812	1.1068	.27116
1.90	.69686	.43936	1.5552	1.5861	1.1083	.27433
1.91	.69379	.43610	1.5677	1.5909	1.1097	.27748
1.92	.69074	.43287	1.5804	1.5957	1.1112	.28061
1.93	.68769	.42967	1.6005	1.5932	1.1126	.28372
1.94	.68465	.42651	1.6062	1.6052	1.1141	.28681
1.95	.68162	.42339	1.6193	1.6099	1.1155	.28989
1.96	.67861	.42030	1.6326	1.6146	1.1170	.29295
1.97	.67561	.41724	1.6461	1.6193	1.1184	.29599
1.98	.67262	.41421	1.6597	1.6239	1.1198	.29902
1.98	.66964	.41121	1.6735	1.6284	1.1213	.30201
2.00	.66667	.40825	1.6875	1.6330	1.1227	.30499
2.01	.66371	.40532	1.7017	1.6375	1.1241	.30796
2.02	.66076	.40241	1.7160	1.6420	1.1255	.31091
2.03	.65783	.39954	1.7305	1.6465	1.1269	.31384
2.04	.65491	.39670	1.7452	1.6509	1.1283	.31675

M	$\frac{T}{T^*}$	$\frac{p}{p^*}$	$\frac{p_o}{p_o^*}$	$\frac{V}{V^*}$	$\frac{F}{F^*}$	$\frac{\lambda_{av} L_{max}}{D}$
2.05	.65200	.39389	1.7600	1.6553	1.1297	.31965
2.06	.64910	.39110	1.7750	1.6597	1.1311	.32253
2.07	.64621	.38834	1.7902	1.6640	1.1325	.32538
2.08	.64333	.38562	1.8056	1.6683	1.1339	.32822
2.09	.64047	.38292	1.8212	1.6726	1.1352	.33104
2.10	.63762	.38024	1.8369	1.6769	1.1366	.33385
2.11	.63478	.37760	1.8528	1.6811	1.1380	.33664
2.12	.63195	.37498	1.8690	1.6853	1.1393	.33940
2.13	.62914	.37239	1.8853	1.6895	1.1407	.34215
2.14	.62633	.36982	1.9018	1.6936	1.1420	.34488
2.15	.62354	.36728	1.9185	1.6977	1.1434	.34760
2.16	.62076	.36476	1.9354	1.7018	1.1447	.35030
2.17	.61799	.36227	1.9525	1.7059	1.1460	.35298
2.18	.61523	.35980	1.9698	1.7099	1.1474	.35564
2.19	.61249	.35736	1.9873	1.7139	1.1487	.35828
2.20	.60976	.35494	2.0050	1.7179	1.1500	.36091
2.21	.60704	.35254	2.0228	1.7219	1.1513	.36352
2.22	.60433	.35017	2.0409	1.7258	1.1526	.36611
2.23	.60163	.34782	2.0592	1.7297	1.1539	.36868
2.24	.59895	.34550	2.0777	1.7336	1.1552	.37124
2.25	.59627	.34319	2.0964	1.7374	1.1565	.37378
2.26	.59361	.34091	2.1154	1.7412	1.1578	.37630
2.27	.59096	.33865	2.1345	1.7450	1.1590	.37881
2.28	.58833	.33641	2.1538	1.7488	1.1603	.38130
2.29	.58570	.33420	2.1733	1.7526	1.1616	.38377
2.30	.58309	.33200	2.1931	1.7563	1.1629	.38623
2.31	.58049	.32983	2.2131	1.7600	1.1641	.38867
2.32	.57790	.32767	2.2333	1.7637	1.1653	.39109
2.33	.57532	.32554	2.2537	1.7673	1.1666	.39350
2.34	.57276	.32342	2.2744	1.7709	1.1678	.39589

M	$\frac{T}{T^*}$	$\frac{p}{p^*}$	$\frac{p_o}{p_o^*}$	$\frac{V}{V^*}$	$\frac{F}{F^*}$	$\frac{\lambda_{av} L_{max}}{D}$
2.35	.57021	.32133	2.2953	1.7745	1.1690	.39826
2.36	.56767	.31925	2.3164	1.7781	1.1703	.40062
2.37	.56514	.31720	2.3377	1.7817	1.1715	.40296
2.38	.56262	.31516	2.3593	1.7852	1.1727	.40528
2.39	.56011	.31314	2.3811	1.7887	1.1739	.40760
2.40	.55762	.31114	2.4031	1.7922	1.1751	.40989
2.41	.55514	.30916	2.4254	1.7956	1.1763	.41216
2.42	.55267	.30720	2.4479	1.7991	1.1775	.41442
2.43	.55021	.30525	2.4706	1.8025	1.1786	.41667
2.44	.54776	.30332	2.4936	1.8059	1.1798	.41891
2.45	.54533	.30141	2.5168	1.8092	1.1810	.42113
2.46	.54291	.29952	2.5403	1.8126	1.1821	.42333
2.47	.54050	.29765	2.5640	1.8159	1.1833	.42551
2.48	.53810	.29579	2.5880	1.8192	1.1844	.42768
2.49	.53571	.29395	2.6122	1.8225	1.1856	.42983
2.50	.53333	.29212	2.6367	1.8257	1.1867	.43197
2.51	.53097	.29031	2.6615	1.8290	1.1879	.43410
2.52	.52862	.28852	2.6865	1.8322	1.1890	.43621
2.53	.52627	.28674	2.7117	1.8354	1.1910	.43831
2.54	.52394	.28498	2.7372	1.8386	1.1912	.44040
2.55	.52163	.28323	2.7630	1.8417	1.1923	.44247
2.56	.51932	.28150	2.7891	1.8448	1.1934	.44452
2.57	.51702	.27978	2.8154	1.8479	1.1945	.44655
2.58	.51474	.27808	2.8420	1.8510	1.1956	.44857
2.59	.51247	.27640	2.8689	1.8541	1.1967	.45059
2.60	.51020	.27473	2.8960	1.8571	1.1978	.45259
2.61	.50795	.27307	2.9234	1.8602	1.1989	.45457
2.62	.50571	.27143	2.9511	1.8632	1.2000	.45654
2.63	.50349	.26980	2.9791	1.8662	1.2011	.45850
2.64	.50127	.26818	3.0074	1.8691	1.2021	.46044

M	$\frac{T}{T^*}$	$\frac{p}{p^*}$	$\frac{p_o}{p_o^*}$	$\frac{V}{V^*}$	$\frac{F}{F^*}$	$\frac{\lambda_{av} L_{max}}{D}$
2.65	.49906	.26658	3.0359	1.8721	1.2031	.46237
2.66	.49687	.26499	3.0647	1.8750	1.2042	.46429
2.67	.49469	.26342	3.0938	1.8779	1.2052	.46619
2.68	.49251	.26186	3.1234	1.8808	1.2062	.46807
2.69	.49035	.26032	3.1530	1.8837	1.2073	.46996
2.70	.48820	.25878	3.1830	1.8865	1.2083	.47182
2.71	.48606	.25726	3.2133	1.8894	1.2093	.47367
2.72	.48393	.25575	3.2440	1.8922	1.2103	.47551
2.73	.48182	.25426	3.2749	1.8950	1.2113	.47734
2.74	.47971	.25278	3.3061	1.8978	1.2123	.47915
2.75	.47761	.25131	3.3376	1.9005	1.2133	.48095
2.76	.47553	.24985	3.3695	1.9032	1.2143	.48274
2.77	.47346	.24840	3.4017	1.9060	1.2153	.48452
2.78	.47139	.24697	3.4342	1.9087	1.2163	.48628
2.79	.46933	.24555	3.4670	1.9114	1.2173	.48803
2.80	.46729	.24414	3.5001	1.9140	1.2182	.48976
2.81	.46526	.24274	3.5336	1.9167	1.2192	.49148
2.82	.46324	.24135	3.5674	1.9193	1.2202	.49321
2.83	.46122	.23997	3.6015	1.9220	1.2211	.49491
2.84	.45922	.23861	3.6359	1.9246	1.2221	.49660
2.85	.45723	.23726	3.6707	1.9271	1.2230	.49828
2.86	.45525	.23592	3.7058	1.9297	1.2240	.49995
2.87	.45328	.23458	3.7413	1.9322	1.2249	.50161
2.88	.45132	.23326	3.7771	1.9348	1.2258	.50326
2.89	.44937	.23196	3.8133	1.9373	1.2268	.50489
2.90	.44743	.23066	3.8498	1.9398	1.2277	.50651
2.91	.44550	.22937	3.8866	1.9423	1.2286	.50812
2.92	.44358	.22809	3.9238	1.9448	1.2295	.50973
2.93	.44167	.22682	3.9614	1.9472	1.2304	.51133
2.94	.43977	.22556	3.9993	1.9497	1.2313	.51291

M	$\frac{T}{T^*}$	$\frac{p}{p^*}$	$\frac{p_o}{p_o^*}$	$\frac{V}{V^*}$	$\frac{F}{F^*}$	$\frac{\lambda_{av} L_{max}}{D}$
2.95	.43788	.22431	4.0376	1.9521	1.2322	.51447
2.96	.43600	.22307	4.0763	1.9545	1.2331	.51603
2.97	.43413	.22185	4.1153	1.9569	1.2340	.51758
2.98	.43226	.22063	4.1547	1.9592	1.2348	.51912
2.99	.43041	.21942	4.1944	1.9616	1.2357	.52064
3.0	.42857	.21822	4.2346	1.9640	1.2366	.52216
3.5	.34783	.16850	6.7896	2.0642	1.2743	.58643
4.0	.28571	.13363	10.719	2.1381	1.3029	.63306
4.5	.23762	.10833	16.562	2.1936	1.3247	.66764
5.0	.20000	.08944	25.000	2.2361	1.3416	.69381
6.0	.14634	.06376	53.180	2.2953	1.3655	.72987
7.0	.11111	.04762	104.14	2.3333	1.3810	.75281
8.0	.08696	.03686	190.11	2.3591	1.3915	.76820
9.0	.06977	.02935	327.19	2.3772	1.3989	.77898
10.0	.05714	.02390	535.94	2.3905	1.4044	.78683
∞	0	0	∞	2.4495	1.4289	.82153

Table-4 One Dimensional Normal Shock Functions.

For a perfect gas with constant specific heat and molecular weight $k = 1.4$.

M_x	M_y	$\frac{p_y}{p_x}$	$\frac{\rho_y}{\rho_x}$	$\frac{T_y}{T_x}$	$\frac{p_{oy}}{p_{ox}}$	$\frac{p_{oy}}{p_x}$
1.00	1.00000	1.00000	1.00000	1.00000	1.00000	1.8929
1.01	.99013	1.02345	1.01669	1.00665	.99999	1.9152
1.02	.98052	1.04713	1.03344	1.01325	.99998	1.9379
1.03	.97115	1.07105	1.05024	1.01981	.99997	1.9610
1.04	.96202	1.09520	1.06709	1.02634	.99994	1.9845
1.05	.95312	1.1196	1.08398	1.03284	.99987	2.0083
1.06	.94444	1.1442	1.10092	1.03931	.99976	2.0325
1.07	.93598	1.1690	1.11790	1.04575	.99962	2.0570
1.08	.92772	1.1941	1.13492	1.05217	.99944	2.0819
1.09	.91965	1.2194	1.15199	1.05856	.99921	2.1072
1.10	.91177	1.2450	1.1691	1.06494	.99892	2.1328
1.11	.90408	1.2708	1.1862	1.07130	.99858	2.1588
1.12	.89656	1.2968	1.2034	1.07764	.99820	2.1851
1.13	.88922	1.3230	1.2206	1.08396	.99776	2.2118
1.14	.88204	1.3495	1.2378	1.09027	.99726	2.2388
1.15	.87502	1.3762	1.2550	1.09657	.99669	2.2661
1.16	.86816	1.4032	1.2723	1.10287	.99605	2.2937
1.17	.86145	1.4304	1.2896	1.10916	.99534	2.3217
1.18	.85488	1.4578	1.3069	1.11544	.99455	2.3499
1.19	.84846	1.4854	1.3243	1.12172	.99371	2.3786
1.20	.84217	1.5133	1.3416	1.1280	.99280	2.4075
1.21	.83601	1.5414	1.3590	1.1343	.99180	2.4367
1.22	.82998	1.5698	1.3764	1.1405	.99073	2.4662
1.23	.82408	1.5984	1.3938	1.1468	.98957	2.4961
1.24	.81830	1.6272	1.4112	1.1531	.98835	2.5263

M_x	M_y	$\frac{p_y}{p_x}$	$\frac{\rho_y}{\rho_x}$	$\frac{T_y}{T_x}$	$\frac{p_{oy}}{p_{ox}}$	$\frac{p_{oy}}{p_x}$
1.25	.81264	1.6562	1.4286	1.1594	.98706	2.5568
1.26	.80709	1.6855	1.4460	1.1657	.98568	2.5876
1.27	.80165	1.7150	1.4634	1.1720	.98422	2.6187
1.28	.79631	1.7448	1.4808	1.1782	.98268	2.6500
1.29	.79108	1.7748	1.4983	1.1846	.98106	2.6816
1.30	.78596	1.8050	1.5157	1.1909	.97935	2.7135
1.31	.78093	1.8354	1.5331	1.1972	.97758	2.7457
1.32	.77600	1.8661	1.5505	1.2035	.97574	2.7783
1.33	.77116	1.8970	1.5680	1.2099	.97382	2.8112
1.34	.76641	1.9282	1.5854	1.2162	.97181	2.8444
1.35	.76175	1.9596	1.6028	1.2226	.96972	2.8778
1.36	.75718	1.9912	1.6202	1.2290	.96756	2.9115
1.37	.75269	2.0230	1.6376	1.2354	.96534	2.9455
1.38	.74828	2.0551	1.6550	1.2418	.96304	2.9798
1.39	.74396	2.0874	1.6723	1.2482	.96065	3.0144
1.40	.73971	2.1200	1.6896	1.2547	.95819	3.0493
1.41	.73554	2.1528	1.7070	1.2612	.95566	3.0844
1.42	.73144	2.1858	1.7243	1.2676	.95306	3.1198
1.43	.72741	2.2190	1.7416	1.2742	.95039	3.1555
1.44	.72345	2.2525	1.7589	1.2807	.94765	3.1915
1.45	.71956	2.2862	1.7761	1.2872	.94483	3.2278
1.46	.71574	2.3202	1.7934	1.2938	.94196	3.2643
1.47	.71198	2.3544	1.8106	1.3004	.93901	3.3011
1.48	.70829	2.3888	1.8278	1.3070	.93600	3.3382
1.49	.70466	2.4234	1.8449	1.3136	.93292	3.3756
1.50	.70109	2.4583	1.8621	1.3202	.92978	3.4133
1.51	.69758	2.4934	1.8792	1.3269	.92658	3.4512
1.52	.69413	2.5288	1.8962	1.3336	.92331	3.4894
1.53	.69073	2.5644	1.9133	1.3403	.91999	3.5279
1.54	.68739	2.6003	1.9303	1.3470	.91662	3.5667

M_x	M_y	$\frac{p_y}{p_x}$	$\frac{\rho_y}{\rho_x}$	$\frac{T_y}{T_x}$	$\frac{p_{oy}}{p_{ox}}$	$\frac{p_{oy}}{p_x}$
1.55	.68410	2.6363	1.9473	1.3538	.91319	3.6058
1.56	.68086	2.6725	1.9643	1.3606	.90970	3.6451
1.57	.67768	2.7090	1.9812	1.3674	.90615	3.6847
1.58	.67455	2.7458	1.9981	1.3742	.90255	3.7245
1.59	.67147	2.7828	2.0149	1.3811	.89889	3.7645
1.60	.66844	2.8201	2.0317	1.3880	.89520	3.8049
1.61	.66545	2.8575	2.0485	1.3949	.89144	3.8456
1.62	.66251	2.8951	2.0652	1.4018	.88764	3.8866
1.63	.65962	2.9330	2.0820	1.4088	.88380	3.9278
1.64	.65677	2.9712	2.0986	1.4158	.87992	3.9693
1.65	.65396	3.0096	2.1152	1.4228	.87598	4.0111
1.66	.65119	3.0482	2.1318	1.4298	.87201	4.0531
1.67	.64847	3.0870	2.1484	1.4369	.86800	4.0954
1.68	.64579	3.1261	2.1649	1.4440	.86396	4.1379
1.69	.64315	3.1654	2.1813	1.4512	.85987	4.1807
1.70	.64055	3.2050	2.1977	1.4583	.85573	4.2238
1.71	.63798	3.2448	2.2141	1.4655	.85155	4.2672
1.72	.63545	3.2848	2.2304	1.4727	.84735	4.3108
1.73	.63296	3.3250	2.2467	1.4800	.84312	4.3547
1.74	.63051	3.3655	2.2629	1.4873	.83886	4.3989
1.75	.62809	3.4062	2.2791	1.4946	.83456	4.4433
1.76	.62570	3.4472	2.2952	1.5019	.83024	4.4880
1.77	.62335	3.4884	2.3113	1.5093	.82589	4.5330
1.78	.62104	3.5298	2.3273	1.5167	.82152	4.5783
1.79	.61875	3.5714	2.3433	1.5241	.81711	4.6233
1.80	.61650	3.6133	2.3592	1.5316	.81268	4.6695
1.81	.61428	3.6554	2.3751	1.5391	.80823	4.7155
1.82	.61209	3.6978	2.3909	1.5466	.80376	4.7618
1.83	.60993	3.7404	2.4067	1.5542	.79926	4.8083
1.84	.60780	3.7832	2.4224	1.5617	.79474	4.8551

M_x	M_y	$\frac{p_y}{p_x}$	$\frac{\rho_y}{\rho_x}$	$\frac{T_y}{T_x}$	$\frac{p_{oy}}{p_{ox}}$	$\frac{p_{oy}}{p_x}$
1.85	.60570	3.8262	2.4381	1.5694	.79021	4.9022
1.86	.60363	3.8695	2.4537	1.5770	.78567	.49498
1.87	.60159	3.9130	2.4693	1.5847	.78112	4.9974
1.88	.59957	3.9568	2.4848	1.5924	.77656	5.0453
1.89	.59758	4.0008	2.5003	1.6001	.77197	5.0934
1.90	.59562	4.0450	2.5157	1.6079	.76735	5.1417
1.91	.59368	4.0894	2.5310	1.6157	.76273	5.1904
1.92	.59177	4.1341	2.5463	1.6236	.75812	5.2394
1.93	.58988	4.1790	2.5615	1.6314	.75347	5.2886
1.94	.58802	4.2242	2.5767	1.6394	.74883	5.3381
1.95	.58618	4.2696	2.5919	1.6473	.74418	5.3878
1.96	.58437	4.3152	2.6070	1.6553	.73954	5.4378
1.97	.58258	4.3610	2.6220	1.6633	.73487	5.4880
1.98	.58081	4.4071	2.6369	1.6713	.73021	5.5385
1.99	.57907	4.4534	2.6518	1.6794	.72554	5.5894
2.00	.57735	4.5000	2.6666	1.6875	.72088	5.6405
2.01	.57565	4.5468	2.6814	1.6956	.71619	5.6918
2.02	.57397	4.5938	2.6962	1.7038	.71152	5.7434
2.03	.57231	4.6411	2.7109	1.7120	.70686	5.7952
2.04	.57068	4.6886	2.7255	1.7203	.70218	5.8473
2.05	.56907	4.7363	2.7400	1.7286	.69752	5.8997
2.06	.56747	4.7842	2.7545	1.7369	.69284	5.9523
2.07	.56589	4.8324	2.7690	1.7452	.68817	6.0052
2.08	.56433	4.8808	2.7834	1.7536	.68351	6.0584
2.09	.56280	4.9295	2.7977	1.7620	.67886	6.1118
2.10	.56128	4.9784	2.8119	1.7704	.67422	6.1655
2.11	.55978	5.0275	2.8216	1.7789	.66957	6.2194
2.12	.55830	5.0768	2.8402	1.7874	.66492	6.2736
2.13	.55683	5.1264	2.8543	1.7960	.66029	6.3280
2.14	.55538	5.1762	2.8683	1.8046	.65567	6.3827

M_x	M_y	$\frac{p_y}{p_x}$	$\frac{\rho_y}{\rho_x}$	$\frac{T_y}{T_x}$	$\frac{p_{oy}}{p_{ox}}$	$\frac{p_{oy}}{p_x}$
2.15	.55395	5.2262	2.8823	1.8132	.65105	6.4377
2.16	.55254	5.2765	2.8962	1.8219	.64644	6.4929
2.17	.55114	5.3270	2.9100	1.8306	.64185	6.5484
2.18	.54976	5.3778	2.9238	1.8393	.63728	6.6042
2.19	.54841	5.4288	2.9376	1.8481	.63270	6.6602
2.20	.54706	5.4800	2.9512	1.8569	.62812	6.7163
2.21	.54572	5.5314	2.9648	1.8657	.62358	6.7730
2.22	.54440	5.5831	2.9783	1.8746	.61905	6.8299
2.23	.54310	5.6350	2.9918	1.8835	.61453	6.8869
2.24	.54182	5.6872	3.0052	1.8924	.61002	6.9442
2.25	.54055	5.7396	3.0186	1.9014	.60554	7.0018
2.26	.53929	5.7922	3.0319	1.9104	.60106	7.0597
2.27	.53805	5.8451	3.0452	1.9194	.59659	7.1178
2.28	.53683	5.8982	3.0584	1.9285	.59214	7.1762
2.29	.53561	5.9515	3.0715	1.9376	.58772	7.2348
2.30	.53441	6.0050	3.0846	1.9468	.58331	7.2937
2.31	.53322	6.0588	3.0976	1.9560	.57891	7.3529
2.32	.53205	6.1128	3.1105	1.9652	.57452	7.4123
2.33	.53089	6.1670	3.1234	1.9745	.57015	7.4720
2.34	.52974	6.2215	3.1362	1.9838	.56580	7.5319
2.35	.52861	6.2762	3.1490	1.9931	.56148	7.5920
2.36	.52749	6.3312	3.1617	2.0025	.55717	7.6524
2.37	.52638	6.3864	3.1743	2.0119	.55288	7.7131
2.38	.52528	6.4418	3.1869	2.0213	.54862	7.7741
2.39	.52419	6.4974	3.1994	2.0308	.54438	7.8354
2.40	.52312	6.5533	3.2119	2.0403	.54015	7.8969
2.41	.52206	6.6094	3.2243	2.0499	.53594	7.9587
2.42	.52100	6.6658	3.2366	2.0595	.53175	8.0207
2.43	.51996	6.7224	3.2489	2.0691	.52758	8.0830
2.44	.51894	6.7792	3.2611	2.0788	.52344	8.1455

M_x	M_y	$\frac{p_y}{p_x}$	$\frac{\rho_y}{\rho_x}$	$\frac{T_y}{T_x}$	$\frac{p_{oy}}{p_{ox}}$	$\frac{p_{oy}}{p_x}$
2.45	.51792	6.8362	3.2733	2.0885	.51932	8.2083
2.46	.51691	6.8935	3.2854	2.0982	.51521	8.2714
2.47	.51592	6.9510	3.2975	2.1080	.51112	8.3347
2.48	.51493	7.0088	3.3095	2.1178	.50706	8.3983
2.49	.51395	7.0668	3.3214	2.1276	.50303	8.4622
2.50	.51299	7.1250	3.3333	2.1375	.49902	8.5262
2.51	.51204	7.1834	3.3451	2.1474	.49502	8.5904
2.52	.51109	7.2421	3.3569	2.1574	.49104	8.6549
2.53	.51015	7.3010	3.3686	2.1674	.48709	8.7198
2.54	.50923	7.3602	3.3802	2.1774	.48317	8.7850
2.55	.50831	7.4196	3.3918	2.1875	.47927	8.8505
2.56	.50740	7.4792	3.4034	2.1976	.47540	8.9162
2.57	.50651	7.5391	3.4149	2.2077	.47155	8.9821
2.58	.50562	7.5992	3.4263	2.2179	.46772	9.0482
2.59	.50474	7.6595	3.4376	2.2281	.46391	9.1146
2.60	.50387	7.7200	3.4489	2.2383	.46012	9.1813
2.61	.50301	7.7808	3.4602	2.2486	.45636	9.2481
2.62	.50216	7.8418	3.4714	2.2589	.45262	9.3154
2.63	.50132	7.9030	3.4825	2.2693	.44891	9.3829
2.64	.50048	7.9645	3.4936	2.2797	.44522	9.4507
2.65	.49965	8.0262	3.5047	2.2901	.44155	9.5187
2.66	.49883	8.0882	3.5157	2.3006	.43791	9.5869
2.67	.49802	8.1504	3.5266	2.3111	.43429	9.6553
2.68	.49722	8.2128	3.5374	2.3217	.43070	9.7241
2.69	.49642	8.2754	3.5482	2.3323	.42713	9.7932
2.70	.49563	8.3383	3.5590	2.3429	.42359	9.8625
2.71	.49485	8.4014	3.5697	2.3536	.42007	9.9320
2.72	.49408	8.4648	3.5803	2.3643	.41657	10.002
2.73	.49332	8.5284	3.5909	2.3750	.41310	10.072
2.74	.49256	8.5922	3.6014	2.3858	.40965	10.142

M_x	M_y	$\frac{p_y}{p_x}$	$\frac{\rho_y}{\rho_x}$	$\frac{T_y}{T_x}$	$\frac{p_{oy}}{p_{ox}}$	$\frac{p_{oy}}{p_x}$
2.75	.49181	8.6562	3.6119	2.3966	.40622	10.212
2.76	.49107	8.7205	3.6224	2.4074	.40282	10.283
2.77	.49033	8.7850	3.6328	2.4183	.39945	10.354
2.78	.48960	8.8497	3.6431	2.4292	.39610	10.426
2.79	.48888	8.9147	3.6533	2.4402	.39276	10.498
2.80	.48817	8.9800	3.6635	2.4512	.38946	10.569
2.81	.48746	9.0454	3.6737	2.4622	.38618	10.641
2.82	.48676	9.1111	3.6838	2.4733	.38293	10.714
2.83	.48607	9.1770	3.6939	2.4844	.37970	10.787
2.84	.48538	9.2432	3.7039	2.4955	.37649	10.860
2.85	.48470	9.3096	3.7139	2.5067	.37330	10.933
2.86	.48402	9.3762	3.7238	2.5179	.37013	11.006
2.87	.48334	9.4431	3.7336	2.5292	.36700	11.080
2.88	.48268	9.5102	3.7434	2.5405	.36389	11.154
2.89	.48203	9.5775	3.7532	2.5518	.36080	11.228
2.90	.48138	9.6450	3.7629	2.5632	.35773	11.302
2.91	.48074	9.7127	3.7725	2.5746	.35469	11.377
2.92	.48010	9.7808	3.7821	2.5860	.35167	11.452
2.93	.47946	9.8491	3.7917	2.5975	.34867	11.527
2.94	.47883	9.9176	3.8012	2.6090	.34570	11.603
2.95	.47821	9.986	3.8106	2.6206	.34275	11.679
2.96	.47760	10.055	3.8200	2.6322	.33982	11.755
2.97	.47699	10.124	3.8294	2.6438	.33692	11.831
2.98	.47638	10.194	3.8387	2.6555	.33404	11.907
2.99	.47578	10.263	3.8479	2.6672	.33118	11.984
3.00	.47519	10.333	3.8571	2.6790	.32834	12.061
3.50	.45115	14.125	4.2608	3.3150	.21295	16.242
4.00	.43496	18.500	4.5714	4.0469	.13876	21.068
4.50	.42355	23.458	4.8119	4.8751	.09170	26.539
5.00	.41523	29.000	5.0000	5.8000	.06172	32.654

M_x	M_y	$\frac{p_y}{p_x}$	$\frac{\rho_y}{\rho_x}$	$\frac{T_y}{T_x}$	$\frac{p_{oy}}{p_{ox}}$	$\frac{p_{oy}}{p_x}$
6.00	.40416	41.833	5.2683	7.941	.02965	46.815
7.00	.39736	57.000	5.4444	10.469	.01535	63.552
8.00	.39289	74.500	5.5652	13.387	.00849	82.865
9.00	.38980	94.333	5.6512	16.693	.00496	104.753
10.00	.38757	116.50	5.7143	20.388	.00304	129.217
∞	.37796	∞	6.000	∞	0	∞

BIBLIOGRAPHY

Anderson, J.D., Jr., Modern Compressible flow, McGraw-Hill, New York, 1982.

John, J.E.A., Gas Dynamics, Allyn and Bacon, Boston, 1984.

Liepmann, H.W. and Roshko, A., Elements of Gas Dynamics, Wiley, New York, 1963.

Shapiro, A.H., Dynamics and Thermodynamics of compressible fluid flow Vol. I, John Wiley and Sons, 1988.

Zucrow, M.J. and Hoffman, J.D., Gas Dynamics Vol. I, Wiley, 1976.

Rathakrishnan, E., Gas Dynamics, Prentice Hall of India, New Delhi, 1995.

Pai, S.I., Introduction to the theory of Compressible flow, Von Nostrand, 1959.

Oswatitsch, K., Gas Dynamics, International text book, 1964.

Rotty, R.M., Introduction to Gas Dynamics, John Wiley and Sons, 1962.

Doneshyar, H., One dimensional compressible flow, Pergamon Press, 1976.

Hussain Zoeb, Gas Dynamics through Problems, Wiley Eastern Ltd., 1989.

Munson, B.R., Young, D.F., Okiishi, T.H., Fundamentals of Fluid Mechanics, John Wiley and Sons 1990.

Chapman, A.J., Walker, W.F., Introductory Gas Dynamics, Holt, Rinehart and Winston, Inc., 1971.

Cambel, A.B., Jennings, B.H., Gas Dynamics, McGraw Hill Book Comp., Inc., 1958.

Hansen, A.G., Fluid Mechanics, John Wiley and Sons 1967.

Cheers, Elements of Compressible flow, John Wiley 1963.

Imrie, B.W., Compressible fluid flow, Butter Worths 1973.

Thompson, P.A. Compressible fluid Dynamics, McGraw-Hill, 1972.

Zucker, Fundamentals of Gas Dynamics, Matrix Publishers, 1977.

Benedict and Sheltz, Handbook of generalized Gas Dynamics, Plenum Press 1966.

Tsien, H.S., Equations of Gas Dynamics, McGraw Hill, 1958.

Yahya, S.M., Fundamentals of compressible flow, Wiley Eastern Limited, 1992.

Yadao, K.L., Fundamentals of Gas Dynamics, Khanna Publishers, 1988.

Yahya, S.M., Elementary Gas Dynamics, Satya Prakashan, 1978.

Purohit, R.K., Preparation of isentropic flow and normal shock tables, (submitted by final year B.E. (Mechanical) Engineering Students, J.N.V. University Jodhpur, Raj., 1996).

INDEX